European Technical Coatings Congress (ETCC 2022)

Coatings Inspirations

AA005567

Krakow, Poland
12-14 July 2022

ISBN: 978-1-7138-7633-5

Printed from e-media with permission by:

Curran Associates, Inc.
57 Morehouse Lane
Red Hook, NY 12571

Polish Association of Chemical Engineers
SITPChem Main Board
ul. Tadeusza Czackiego 3/5
00-043 Warsaw

Phone: +48 22 826 78 96

https://sitpchem.org.pl/polish-association-of-chemical-engineers-sitpchem/

Additional copies of this publication are available from:

Curran Associates, Inc.
57 Morehouse Lane
Red Hook, NY 12571 USA
Phone: 845-758-0400
Fax: 845-758-2633
Email: curran@proceedings.com
Web: www.proceedings.com

TABLE OF CONTENTS

PLENARY

SESSION 1

SESSION 2

SESSION 3

SESSION 4

SESSION 5

SESSION 6

SESSION 7

SESSION 8

SESSION 13

SESSION 14

SESSION 15

SESSION 16

SESSION 17

SESSION 18

SESSION 19

SESSION 20

SESSION 21

SESSION 22

SESSION 23

SESSION 24

SESSION 25

SESSION 26

SESSION 31

SESSION 32

SESSION 33

SESSION 34

SESSION 35

SESSION 36

POSTER SESSION

12-14 July 2022 Krakow, Poland

ETCC 2022 | Coatings Inspirations

EUROPEAN TECHNICAL COATINGS CONGRESS

Since 1950 www.etcc2022.org

Creating Value through Innovation and Collaboration – Andre van Linden (AkzoNobel R&D)

In a world where the customer expects the paint industry to deliver more and better than ever, we will have to go even further. Beyond expectation by delivering much further going functionality, improved durability and uses on new substrates for paints. Where paint historically was intended for protection and beautifying the subject it is now often expected to add something to the substrate or environment where it is applied. By going beyond imagination this will be into completely new functionalities for paints but also in the way it is offered to customers. By thinking beyond generations the sustainability is to be taken into account and by noticing and comparing what is done in carbon footprint gains by reducing, reusing / recycling and renewing the products and raw materials in our portfolio it can be seen where the biggest steps can still be made. By combining the new chemical possibilities with use in new functionalities it is clear that only research and innovation where the whole chain of collaborations from start-ups to universities is used is leading to the new opportunities we are now seeing (or will soon see) in the market and to the sustainability objectives that are promised (50% carbon footprint reduction by 2030).

New solutions for Coatings: Taking you on a journey from current developments to future trends and beyond

Rüdiger Mertsch, Annett Halbhuber, Sascha Herrwerth, Matthias Kremser

Evonik Resource Efficiency GmbH, Hanau (Germany)

Developments in the coatings industry are often driven by regulatory needs and a trend to more **sustainability**. In addition to **VOC reduction** and the demand for high solid and waterborne coating formulations also more **durable** and even **functional coatings** [1] are required.

Waterborne formulations are already quite common in decorative paints and for do-it-yourself (DIY) applications. Nevertheless, this market as well as others face a diversity of challenges like eco-labeling, biocide restrictions and many more. A major market pull is now coming from the industrial coatings sector towards adopting waterborne systems. The Chinese industrial coatings sector (e.g. container coatings market) is a modern-day example of how fast a market may need to switch to waterborne technology solutions on short notice, mainly driven by governmental restrictions. In response, the coatings manufacturer and the raw material suppliers have to react with new solutions for the waterborne market.

At the same time, it is mandated that the coating process itself becomes faster, less complex and more cost efficient. **Energy saving** and **resource efficiency** are monitored by "cradle to grave" **life cycle analysis** [2] including e.g. **CO_2 footprint, VOC emission** and the **primary energy demand** of the coating raw materials as well as the total production process.

The need of more **durable, resource efficient** [3] and even **functional coatings** [1] also requires the d*evelopment* of new binder and/or hardener systems. These enhanced properties can be achieved through smart **hybridization** [4] of binder systems aligned with the combination of different curing mechanisms (e.g. UV-LED curing, Michael-addition, NISO etc.). As an example, the versatile hybridization technology of organic and silicone/silane chemistry will be discussed in detail.

This presentation will give an overview of various methods and technologies complying with the desire for **functional coatings** are presented. These include: new **biocide-free technologies** [5] for **anti-microbial** [6] and **foul-release** coatings driven by biocide restrictions, as well as topics/technologies for **thermal insulating** coatings. It will also detail successfully implemented processes in today's technology and their further development.

Finally, a short outlook on upcoming **smart coatings** such as the reversible change of surface properties (color, hydrophilic <> hydrophobic), sensor coatings and electro-conductive coatings will be addressed.

References:

[1] V. Stenzel, N. Rehfeld: *Functional Coatings,* Vincentz Network, Hannover, Germany, **2011** ISBN: **978-3-86630-856-5**

[2] W. Klöpffer, B. Grahl: *Life Cycle Assessment (LCA): A Guide to Best Practice*, Wiley-VCH, Weinheim, Germany, **2014**. ISBN: **978-3-527-32986-1**

[3] B. Bröskamp, S. Herrwerth, F. Böss, C. Cova-Hög, *Resource-efficient corrosion protection*, European Coatings Journal, **2015**, 3, 84-87

[4] W. Heilen, S. Herrwerth: *Silicone resins and their combinations*, Vincentz Network, Hannover, Germany, **2014**. ISBN: **978-3-86630-828-2**

[5] A. Eichner, T. Holzmann, D.B. Eckl, F. Zeman, M. Koller, M. Huber, S. Pemmerl, W. Schneider-Brachert, W. Bäumler*: Novel photodynamic coating reduces the bioburden on near-patient surfaces thereby reducing the risk for onward pathogen transmission: a field study in two hospitals*, Journal of Hospital Infection, **2020**, 104, 85-91

[6] C. Adlhart, J. Verran, N.F. Azevedo, H. Olmez, M.M. Keinänen-Toivola, I. Gouveia, L.F. Melo, F. Crijns*: Surface Modifications for Antimicrobial Effects in the Healthcare Setting: A Critical Overview*, Journal of Hospital Infection, **2018**, 3, 239-249

What is the Appropriate Response of the Coatings Industry with Regard to the Ongoing Process of Reclassification of Compounds ?

Thomas May

May-Ienstein UGS Beratung, Bremen / Germany

In the times before implementation of REACH and CLP, the classification and labelling of substances included, to a certain extent, a combined description of hazards and risks. Intrinsic hazard profiles were expressed by means of risk phrases. And many pieces of occupational health and environmental legislation only took these risk phrases into account to derive practical consequences in the sense of restrictions and limitations. The introduction of REACH and CLP is also a response to the long lasting request by the chemical industry to arrive at a clearer distinction of hazard and risk. CLP now explicitly addresses intrinsic hazards by hazard statements whereas REACH urges manufacturers and users of chemicals to consider exposure and associated risks. A risk is constituted by a combination of hazards and specific conditions of use (e.g. concentration, interaction, duration, process design, setting, ventilation, risk management measures, protection equipment).

Comprehensive substance dossiers according REACH, changed criteria under CLP and political interventions resulted in a reclassification of multiple substances with high relevance for the coatings industry. Especially reactive compounds are at risk to undergo classification changes in the direction of seemingly higher acute toxicity, reprotoxicity, specific target organ toxicity, sensitization, irritation, suspected CMR properties and aquatic toxicity. Whilst in former days it was in most cases viable to replace hazardous compounds by less hazardous alternatives, this escape is becoming more and more unrealistic under the precondition that coatings still need to meet high and even growing expectations with regard to quality and performance. Thus the coatings is expected to assess and communicate more on the conditions of safe use for their products. And this needs to cover the whole life cycle including service life of coated articles and recycling after the end of the service life.

Sector organisations are supporting this transition process by multiple activities. However, quite many manufacturers will still need to enhance their respective services.

References:

[1] Verband der deutschen Lack- und Druckfarbenindustrie e.V. (VdL), Auswirkungen der Einstufung eines Feststoffs als karzinogen, keimzellmutagen oder reproduktionstoxisch im Bereich des Umweltschutzes, der Anlagensicherheit und der Arbeitssicherheit, Frankfurt/Main, Januar 2020 (German paint and printing ink manufacturers association, Consequences of classification of a non-volatile compound as carcinogen, mutagen or reprotoxic in the area of environmental protection, installation safety and occupational safety, Frankfurt/Main, January 2020

Accepted for oral presentation at ETCC 2022

Powder Paints: measure and model the development of appearance

Martin Bosma, Pengcheng Yang, Enrico Rensen

Allnex Corporate Innovation Group, Nieuwe kanaal 7-N, 6709 PA Wageningen, The Netherlands, martin.bosma@allnex.com

Abstract

A key area driving developments in the powder coatings industry today is that of improved energy efficiency through reduction of cure temperature. The ability to cure at lower temperatures offers the dual benefit of both reduced energy use on application and broadening the appeal of powder coatings to include temperature sensitive substrates, such as MDF and plastics. One of the main challenges for low temperature curing powders is to have the ability to sufficiently flow and level at the reduced cure temperature in order to obtain a smooth final appearance while keeping good mechanical properties and powder stability.

In the powder coating industry, the appearance of coatings is typically determined by visual comparison of the appearance of the test panel with those of a standardized set of powder coating panels from the Powder Coating Institute (PCI). This PCI set contains ten different panels of which the appearance ranges from very bad (PCI-1) and gradually improves to very good (PCI-10). The quality of the appearance of a powder coating under study is graded by giving it the PCI panel number its appearance is closest to. It is readily understood that this assessment is highly subjective and may unwantedly vary with variations in lighting conditions, viewing distance and colour of the powder coating.

We have recently developed an optical measuring technique (SVP) that allows the quantitative determination of the appearance of glossy coatings. The results of the SVP can be used to calculate the PCI number in an objective manner. Strong benefit of the SVP method is that it can also be used to study the development of the PCI during the curing process of powder coatings.

This paper describes the results of a recent study where the SVP method was used to identify the dominant factors determining the final appearance of powder coatings. Based on experimental results for among others model powder paints we derived a quantitative model for the PCI appearance of powder paints as a function of powder paint properties and application conditions. The insights obtained were used to facilitate and support development of powder coatings that can be cured at low temperatures while still resulting in good PCI rankings.

Introduction of silver particles in cataphoretic coatings for smart surfaces.

S. Rossi, M. Calovi, F. Russo
Department of Industrial Engineering, University of Trento, Trento, (Italy),
stefano.rossi@unitn.it

The Covid emergency highlighted the necessity to have surfaces with an antimicrobial effect. This is important not only in relation to this latest pandemic but also to numerous other infections that can be transmitted from contaminated surfaces.

The antibacterial efficacy of silver particles has been demonstrated in the literature.

The aim of this work is to evaluate the possibility of adding silver particles in organic coatings deposited by cataphoretic method in order to obtain smart coatings without losing the protective properties.

A commercial pigmentless acrylic cataphoretic bath was used so as not to influence or shield the effect produced by the presence of silver. Carbon steel was used as substrate. The silver was introduced directly into the cataphoretic bath in the form of colloidal silver in a composition of 0.05 and 0.1wt.%. The coating was deposited by application of 60 V for 2 minutes. A curing treatment at 140 °C for 45 minutes was carried out.

A microstructure analysis of the deposited layer was carried out to highlight the presence and the distribution of silver particles. The protective performances of the samples coated with and without silver were evaluated by exposure in a salt spray chamber (ASTM B117) for 500 hours and by electrochemical impedance spectroscopy measurements in 3.5wt. %NaCl solution. The samples were finally exposed to UV-B radiation for 500 h, evaluating the color change of the coatings.

Selective modification of unprotected carbohydrates for cross-linking in waterborne coatings

S.C. Massmann, M.D. Witte, A.J. Minnaard, W.R. Browne, as well as K. van den Bergen (AkzoNobel) and D.J. van Dijk (BASF)

Stratingh institute of chemistry, Groningen (the Netherlands), e-mail: s.c.massmann@rug.nl

Coatings for paints, inks and lacquers on environment-friendly and non-toxic basis are highly desired. Therefore, I am working on the novel functionalization of bio-polymers such as carbohydrates (oligo- and polysaccharides). The group of Prof. Minnaard already showed the selective oxidation of α-methylglucose at the C3 position with a palladium neocuproine catalyst and *p*-benzoquinone in methanol in high yield. This method was demonstrated in the end term oxidation of 1,4-linked glucans up to the 7-mer as well. Through this carbonyl moiety within the carbohydrates, further functionalization is much more accessible and has a wide spectrum of possibilities. During my PhD, I aimed to oxidize carbohydrates and especially polysaccharides with the same selectivity but in a sustainable and greener way. Many screenings were already done focusing on the one hand on stabilizing the Pd catalyst, which has the tendency to decompose due to a hydrogen abstraction from the ligand during the oxidation cycle, and on the other hand on the susceptibility for C3-selective oxidation of different oligo- and polysaccharides.

References:

[1] Ho, W.C., Chung, K., Ingram, A.J., Waymouth, R.M., Pd-catalyzed Aerobic Oxidation Reactions: Strategies To Increase Catalyst Lifetimes. *J. Am. Chem. Soc.*140, 748-757, (2018)

[2] Eisink, N.N.H.M., Witte, M.D., Minnaard, A.J., Regioselective Carbohydrate Oxidations: A Nuclear Magnetic Resonance (NMR) Study on Selectivity, Rate, and Side-Product Formation. *ACS Catal.* 7, 1438-1445, (2017)

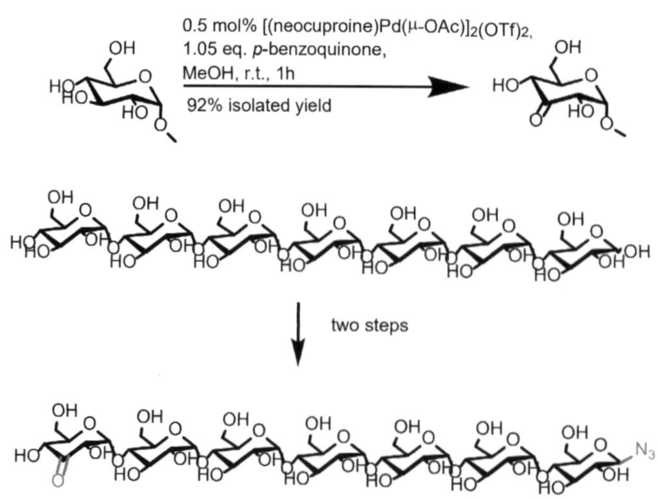

New Strategies for Coatings 2020: Controlled Polymer Synthesis based on NIR Technologies

Ceren Kütahya[1)], Nicolai Meckbach[1)], Zhijun Chen[2)], Bernd Strehmel[1)]

[1)]Niederrhein University of Applied Sciences, Faculty of Chemistry and Institute for Coatings and Surface Chemistry, Adlerstr. 1, D-47798 Krefeld, Germany
[2)]Northeast Forestry University, Key Laboratory of Bio-based Materials, Hexing Road 26, 150040, Harbin, China

In macromolecular and material science, two particular processes utilizing copper catalysts, namely Atom Transfer Radical Polymerization (ATRP) and Copper Catalyzed Azide-Alkyne Cycloaddition Reactions (CuAAC), have intensely influenced the research strategies facilitating fabrication of a wide range of polymers with well-defined structures. Many radical photoinitiators or photosensitizers were reported to be powerful promoters for in situ generation of Cu(I) complexes by the reduction of the corresponding Cu(II) conjugates as a green reaction route. This also have an impact for synthesis in coating sciences including traditional free radical photopolymerization as well[1]. Light as a reagent and tool has also received interest in polymer synthesis.
A new photoinitiating system using NIR sensitizers as successful reducing agents have been suggested for the reduction of Cu(II) species to Cu(I). The produced Cu(I) ions are ideal catalysts for both CuAAC and ATRP reactions. Among the polymethines investigated with cationic and zwitterionic, only zwitterionic carrying a barbital group exhibited sensitization activity under NIR light at room temperature resulting in the formation of polymers with controlled molecular weight characteristics and excellent chain end functionality[2].
Alternatively, it also showed the similar photochemical strategy using carbon dots (**CD**s) as green photosensitizer in photo-ATRP protocol. The use of these cheap and green materials for photopolymerization is successfully studied. These preliminary studies were pursued with blue light LEDs while correct research in the lab extends their exploration with red and NIR light as well[3].
This has additionally worked well with CuAAC using a polymer functionalized at the end with azide and an alkyne. The photomediated click-chemistry based on NIR-sensitizers works well upon exposure with NIR-LEDs at 790 nm[4]. In addition, more studies have been in progress to apply this system for multifunctional azide and multifunctional alkyne monomers carrying the respective alkyne and azide structures. This might be interesting for prospective use in coating applications. In addition, these multi-functional azides and alkynes synthesized can be employed in such NIR-sensitized systems to synthesize polymer networks with homogeneous crosslinking density.

1. Kocaarslan, A.; Kütahya, C.; Keil, D.; Yagci, Y.; Strehmel, B., Near-IR and UV-LED Sensitized Photopolymerization with Onium Salts Comprising Anions of Different Nucleophilicities. *ChemPhotoChem* **2019,** *3* (11), 1127-1132.
2. Kütahya, C.; Schmitz, C.; Strehmel, V.; Yagci, Y.; Strehmel, B., Near-Infrared Sensitized Photoinduced Atom-Transfer Radical Polymerization (ATRP) with a Copper(II) Catalyst Concentration in the ppm Range. *Angew. Chem., Int. Ed.* **2018,** *57* (26), 7898-7902.
3. Kütahya, C.; Wang, P.; Li, S.; Liu, S.; Li, J.; Chen, Z.; Strehmel, B., Carbon Dots as Promising Green Photocatalyst for Free Radical and ATRP-Based Radical Photopolymerization with Blue LEDs. Angewandte Chemie International Edition **2020**, in press, doi: https://doi.org/10.1002/anie.201912343.
4. Kütahya, C.; Yagci, Y.; Strehmel, B., Near-Infrared Photoinduced Copper-Catalyzed Azide-Alkyne Click Chemistry with a Cyanine Comprising a Barbiturate Group. *ChemPhotoChem* **2019,** *3* (11), 1180-1186.

Synthesis of Biomass Derived Monomers for Coating and Polymer Applications

Mukund P. Sibi

Department of Chemistry and Biochemistry, North Dakota State University, Fargo, ND 58108 USA

Approximately 96% of the organic chemicals used today are derived from fossil fuels. However, fossil fuels are a nonrenewable resource with a finite quantity available. With an oil supply that is likely to be depleted in the next 100 years, it is essential to discover alternative renewable resources to produce the chemicals that our world relies on. One viable option is the use of renewable biomass as a chemical feedstock.

The development of green chemical methods for the synthesis of novel monomers for polymer applications has received intense scrutiny in the past two decades. Furthermore, the use of bio-based feedstocks for monomer synthesis has become important due to the projected depletion of fossil fuels in the near future.

Our group's research is focused on the development of novel methods for the conversion of renewable resources to feedstock chemicals for use in polymer synthesis.[1] In our work we have used three important renewable materials (1) oil seeds,[2] (2) cellulose, and (3) lignin.[3] In particular, we are investigating novel methodologies for the synthesis of furanic monomers and their application in polymer synthesis. Our group has synthesized a variety of diols,[4] bis carbamates, diacids,[5] diamines,[6] and dialdehydes.[7] The utility of these monomers in polymer and coating applications will be discussed.

1. Serum, E. M.; Sutton, C. A.; Renner, A. C.; Dawn, D.; Sibi, M. P. New AB Type Monomers from Lignocellulosic Biomass. *Pure Appl. Chem.*, 91(3), 389-396 (2019)
2. Kalita, D.; Kalita, J.; Tarnavchyk, I.; Sibi, M. P.; Moser, B. R. Webster, D. C. Chisholm, B. J. Biobased Poly(vinyl ether)s Derived from Soybean Oil, Linseed Oil, and Camelina Oil: Synthesis, Characterization, and Properties of Crosslinked Networks and Surface Coatings. *Progress in Organic Coatings, 125,* 453-462 (2018).
3. Yu, A. Z.; Serum. E. M.; Renner, A. C.; Sahouani, J. M.; Sibi, M. P.; Webster, D. C. Renewable Reactive Diluents as Practical Styrene Replacements in Bio-based Vinyl Ester Thermosets. *ACS Sustainable Chem. Eng.*, 6(10), 12586-12592 (2018)
4. Vijjamarri, S.; Streed, S.; Serum, E. M.; Sibi, M. P.; Du, G. Polymers from Bio-derived Resources: Synthesis of Poly(silylether)s from Furan Derivatives Catalyzed by a Salen–Mn(V) Complex. *ACS Sustainable Chem. Eng.*, 6 (10), 2491-2497 (2018)
5. Serum, E.M.; Sermadurai, S.; Zimmermann, N.; Sibi, M. P. Valorization of 2,5-Furandicarboxylic Acid. Diels-Alder Reactions with Benzyne. *Green Chem.* 20, 1448-1454 (2018)
6. Samanta, S.; Sermadurai, S.; Bahr, J.; Wicramaratne, D. S.; Sibi, M. P.; Chisholm, B. J. Synthesis and Characterization of Polyurethane Networks Derived from Soybean Oil-Based Cyclic Carbonates and Bio-Based Diamines. *ACS Sustainable Chem. Eng.*, 6551-6561 (2018)
7. Silbert, Samantha D.; Serum, Eric M.; La Scala, John J.; Sibi, Mukund P.; Webster, Dean C. *ACS Sustainable Chem. Eng.*, 7(24), 19621-19630 (2019)

Quick curing mechanisms for architectural all-season coatings

Ivan Cabrera, Immanuel Willerich, Jochen Gattermayer, Thomas Wolf

BASF SE, Ludwigshafen (Germany)

Achieving a balance between quick drying of a paint film, i.e. early humidity resistance, and sufficient open time to re-work the coating for an optimized aesthetic aspect is a challenge for the architectural coatings industry. Current approaches usually favour one of the above aspects for the benefit of the other, thus often leading to 2k systems with a reduced pot-life or separate formulations for winter and summer season.

In this contribution, the mechanisms of different quick curing concepts will be compared comprising traditional systems using either inhibited coagulants that are triggered during paint application or two-component systems that are combined just prior to paint application. Laboratory test methods for screening of quick curing ability of architectural paints and analytical methods like electrophoretic mobility or dynamic light scattering are combined to understand the connection between performance and colloidal properties.

Furthermore, a novel dispersion binder technology for architectural pains is highlighted that differs from traditional quick curing systems by not relying on the evaporation of volatile bases during the drying process. This results in significantly reduced odour compared to many conventional quick-drying systems. Examples show that a balance between early rain resistance and open time under critical temperature and humidity conditions can be achieved without compromising on other performance properties.

MAXIMUM SUSTATAINIBILITY

Huzaifa Matawala, (Paint Recyclist)

Regent Paints Inc, New York, USA, huzaifa@regentpaintsusa.com

We have undoubtedly entered into an era of Abundance. We have not only enhanced our lifespans but also enriched it with unlimited accessories. Such comfort sometimes comes with exploitation of resources. We are in need for Sustainable culture within the industry that covers the unused items back into the cycle. Hence contemplating waste creation in different stages and putting such items to use instead of landfills and burial.

Development and continuous changes are eccentric unavoidable occurrence in Paint Industry. With every such action of chemistry, there are consequences in the inventory levels and waste creations. Generation of obsolete stocks, factory waste, customer returns, expired items, damaged products are unavoidable everyday events. It is in best interest of the industry that such items are reutilized back in production.

We are paint manufacturing company in UAE and INDIA, and recyclers of paints in USA. We are able to recycle and reuse all the non-salable and non-usable back in the production line. Several articles written by me and on me in several magazines explain the chemical processes and the model how we work to reduce carbon footprint of the industry. With a background of reutilizing over 14 million gallons of waste. When spread over the ground can de-fertile land size of approximately 60000 square miles. Land size almost half the size Poland. We have created special coatings for alkyd, acrylic and bituminous formulations that accept anywhere from 30% to 90% in the batches. With a vision to initiate a process to recognize a fact that oil paints have several important ingredients that is better utilized as paints. By burning them into fuel we burn the resins, premium solvents, wax, metallic driers, pigments, additives, other chemicals, etc. When we burn these ingredients, we need to understand the cost of creating a pigment of several grade takes a very intense amount of power and fuel. It also consists of the rarest chemical and metals which is extracted or created with tremendous amount of Utilization of resources.

I had an opportunity to participate in Bio-based Summits held in Bangkok and Dallas in previous 3 months. Due to the amazing work and unique solutions, I'm also honored to chair these events and lead the group discussions.

I would like to present this revolutionary technical and ecological presentation covering all the above in details using the data and graphics from my articles and formulations published earlier in the USA, European, Australian magazines. and elaborated commitment we have entered into with environment on behalf of the Coating Industry.

http://digital.bnpmedia.com/publication/?i=507393&ver=html5&p=50#{"page":50,"issue_id":507393}
https://www.paint.org/article/a-second-look-at-paint-recycling/
https://www.chemarc.com/content/huzaifa-matawala-advanced-paint-recycler-leading-the-way-to-global-reuse-of-paints/5bc71e40c59d694926dd2bdc?in=chemarc%2Fviews%2Fexpert_corner

Chitosan-modified polymeric material for wastewater treatment

George Z. Kyzas
Department of Chemistry, International Hellenic University, Kavala, Greece
kyzas@chem.ihu.gr

One of the most important decontamination techniques is considered to be adsorption. It is fast, simple, low-cost with many opportunities to modify the initial materials after appropriate synthesis routes etc. Numerous adsorbent materials are prepared the last years having as ultimate scope to remove some pollutants especially from contaminated waters (effluents originated from industries). But the composition of each type of effluents is varying. Dyes, heavy metals, pharmaceutical compounds are some major components of industrial wastewaters. In the last years, polymers attracted many researchers employing with adsorption technique, because of their excellent properties. Researchers proceeded with that and achieved even more effective polymeric adsorbents making some modification to initial polymeric chains in order to enhance the adsorption capacity and improve the rigidity, preparing many polymeric composites. The present lecture summarizes the application of a synthesized cross-linked and grafted zwitterionic chitosan derivative to the removal of a mixture of five anti-inflammatory compounds from aqueous matrix. An effective bio-based adsorptive material was synthesized using 2-(methacryloyloxy)ethyl]dimethyl-(3-sulfopropyl)ammonium hydroxide as monomer. According to the functional groups of derivative and adsorption efficiency of chitosan (the polymeric substrate used), results indicated a high adsorption ability of tested material for disposal of the certain remedies from aqueous systems. Parameters like the pH of the solution, the initial concentration of the compounds, the temperature and the contact time between the novel, bio-based adsorptive material and the target compounds proved to affect the process. The optimum pH appears to be pH=4 and adsorption was ascribed to various electrostatic interactions that may take place between the adsorptive material and the drug affected by the functional groups, and the substituents of molecules. For adsorption isotherms, it was found that the Langmuir equation provided a better fitting than the Freundlich equation. Increase of temperature favours the adsorption capacity of the new material while 180 min were found to be the time required for the process to be completed.

Acknowledgment: The financial support received for this study from the Greek Ministry of Development and Investments (General Secretariat for Research and Technology) through the research project "Intergovernmental International Scientific and Technological Innovation-Cooperation. Joint declaration of Science and Technology Cooperation between China and Greece" with the topic "Development of monitoring and removal strategies of emerging micro-pollutants in wastewaters" (Grant no: T7DKI-00220) and it is gratefully acknowledged.

The impact of weather conditions on biocides in paints

Ute Schoknecht

Bundesanstalt für Materialforschung und -prüfung (BAM), Berlin, Germany

ute.schoknecht@bam.de

Keywords: paints, biocides, transformation, emission

Aim. The European biocidal products regulation requires environmental risk assessments for biocidal products under service conditions. This is only possible, if processes that lead to biocide emissions into the environment are understood and can be reasonably predicted.

Actual emissions due to leaching result from different processes that are affected by material properties and environmental conditions. Transformation of biocides can affect emissions considerably.

Leaching and possible transformation of the film preservatives carbendazim, diuron, OIT and terbutryn in paints was observed under laboratory and field conditons in order to investigate the influence of different factors, e.g. water contact and radiation on biocide emissions. The paints contained either white titanium dioxide or a red iron oxide pigment to check whether pigments affect leaching processes.

Methods. Test specimens of a water-based styrene acrylic dispersion paint on glass were exposed to UVA-radiation in a UV weathering device and/or exposed to water contact by immersion events on the basis of the European standard EN 16105. Plywood panels were coated and exposed to natural weather conditions in a semi-field experiment. Concentrations of biocides and transformation products were analysed in leachates from laboratory tests, runoff samples and remaining coatings.

Results. The impact of weathering conditions on transformation processes differs for the investigated active substances. Water contact, radiation as well as pigments in the paints determine the pattern and amount of transformation products on coatings, leachates and runoff water. Observations from laboratory and field experiments were compared. It proved that results from laboratory tests support understanding of data from field experiments.

Detailed knowledge on the fate of active substances under environmental conditions supports manufacturers to develop safe applications of film preservatives and authorities to evaluate its environmental impact.

References:

[1] Bollmann, U.E., Minelgaite, G., Schlüsener, M., Ternes, T., Vollertsen, J., Bester, K. Leaching of terbutryn and its photodegradation products from artificial walls under natural weather conditions. *Environ Sci Technol.* 50, 4289-4295, (2016)

[2] Bollmann, U.E., Minelgaite, G., Schlüsener, M., Ternes, T.A., Vollertsen, J., Bester, K. Photodegration of octylisothiazolinone and semi-field emissions from façade coatings. *Scientific Reports* 7:41501, 1 - 9, (2017)

[3] Urbanczyk, M.M., Bester, K., Borho, N., Schoknecht, U., Bollmann, U.E. Influence of pigments on phototransformation of biocides in paints. *J Hazard Mater.* 3364, 125 – 133, (2019)

Laboratory Weathering with H_2O_2 Spraying

Artur Schönlein[1], Shigeo Suga[2]

[1]RAWS Rüsselsheim (Germany), [2]Suga Test Instruments Co., Ltd. Tokyo (Japan)

Building materials are to be maintained for 30 or more than 50 years. This makes a constant development and adaptation to climatic conditions necessary. In order to keep pace with the development of new materials, there is a constant search for ways to shorten current laboratory testing times. The H_2O_2 spray test proves to be a good accelerating alternative, with a good correlation to outdoor weathering observed. [1]

The aim of a laboratory weathering test is the best possible and accelerated simulation of a material degradation process, which is similarly caused by natural weather factors. Pigments are an important component of paints. They cover the coated material and are crucial for the color impression. Furthermore, pigments provide better resistance of the paint color to UV radiation. Many paints use the white pigment titanium dioxide. [2] It is used in white color and as a base for many more colors.

During outdoor weathering, a shrinkage of paint surfaces due to delamination of titanium dioxide particles and binder degradation is observed. This photocatalytic-based effect can be accelerated by simulating a laboratory exposure test with additional H_2O_2 spraying. [1]

The special xenon weathering test method with an additional H_2O_2 spraying system is explained and test results are discussed.

References

[1] Shun Saito, (2017). Super accelerated weathering test evaluation technology of fluorine resin paint. Suga Weathering Academic Lecture, 2017 65.

[2] Jochen Winkler, (2013). Titandioxid. *Vincentz Network, Hanover, Germany.*

Prediction of protective properties of coatings based on glass transition temperature value

S. Jurczyk

The Łukasiewicz Research Network – Institute for Engineering of Polymer Materials and Dyes, Toruń (Poand) e-mail: s.jurczyk@impib.pl

Dynamic mechanical analysis (DMA) was employed in order to evaluate viscoelastic properties of paint systems. The main value determined by DMA method was glass transition temperature (T_g). The examination objects were acrylic paint systems containing with polyurethane interlayer and acrylic paint system without interlayer. Also were tested coating systems consisted of epoxy primers – with different types of hardeners and anticorrosive pigments, the epoxy intermediate and the polyurethane topcoat crosslinked with hexamethylene diisocyanate (HDI). The evaluations were performed for coatings after neutral salt spray tests and also after UV tests.

The DMA evaluations, performed in shear mode, revealed influences of type and time of exposition on properties of testes paint systems. The glass transition temperature was changed with increasing of aging exposure period. These behaviour indicated hardening of paint systems. T_g value was significant for storage modulus values also. Paints systems were more rigid consistently with time exposure period. For coating systems consisted of epoxy primers good correlation between T_g and protective properties in salt chamber and in changing conditions was visible. An increase in cross-linking frequently leads to a decreased value of *tan δ* and a broader peak, as well as to higher T_g.

Presented results are a confirmation of practical application of DMA for control of coatings formation process and prediction of protective properties. Determination of Tg value and observation relaxation occur glass transition are a very useful tools for estimation of properties paint system/top coat.

References:

[1] Mafi, R., Mirabedini, S.M., Attar, M.M., Moradian, S., Cure characterization of epoxy and polyester clear powder coatings using Differential Scanning Calorimetry (DSC) and Dynamic Mechanical Analysis (DMTA). *Progress in Organic Coatings*. 54 (3), 164 - 169, (2005)

[2] Groenevolt, M., Name4, Y. Highly scratch resistant coatings for automotive applications. *Progress in Organic Coatings*. 61 (2-4), 106 – 109, (2008)

[3] Hu, J., Li, X., Gao, J., Zhao, Q., UV aging characterization of epoxy varnish coated steel upon exposure to artificial weathering environment. *Materials & Design*. 30 (5), 1542 – 1547, (2009)

Scratch/Mar, Surface Structure and Rheology Measurements

Michael Osterhold

Consultant and Lecturer, Bochum, (Germany), michael.osterhold@t-online.de

This contribution yields a brief review of intensively discussed topics over the years in the field of physical characterization of coatings with focus on: scratch resistance, surface structure and rheology measurements.

Scratch/mar is one of the most important physical properties to characterize the mechanical quality of a coating system. The brushes and dirt in a car wash, for example, produce scratches measuring only a few micrometers in width and up to several hundred nanometers in depth. With this background several measuring methods have been discussed over the last 15 to 20 years in the automotive and coatings industry. Procedures that create a single scratch have been developed and improved at the end of the 1990s (micro or nano scratch method). These methods are different from more practically oriented procedures that are based on relatively simple methods to try to test or even come close to reality (e.g. car wash brush method). The methods were reviewed briefly considering also other physical properties (e.g. cross-linking density). [1]

In addition to color, effect and gloss, the visual impression of a painted surface is influenced especially by the surface structure (leveling, waviness, orange peel, appearance). To characterize the structure of a surface, mainly two different measuring methods (profilometry and 'wave-scan') have been established in the automotive and coatings industry over the years. For example, the mechanical profilometry allows detailed investigations concerning the surface topography from substrate to topcoat. Basic relations and relevant application examples from the areas 'metal, plastics, coating', investigated with profilometry and 'wave-scan' in the course of time, will be summarized. [2]

Many application and technical properties of coatings are influenced by their flow behavior. High product quality can only be guaranteed through an exact knowledge of the rheological behavior of the coating and the used raw materials, respectively. In view of the increasing use of waterborne systems flow anomalies as thixotropy, yield points or also viscoelastic behavior can be observed more often. Yield point and thixotropy influence important materials properties as storage stability, pumping behavior or leveling and flowing. Against this background the measuring possibilities of characterizing the rheological properties with rotational rheometers concerning yield point and thixotropy will be presented. [3]

Keywords: scratch resistance, surface structure, rheology measurements

References:

[1] Osterhold, M. Marking time. *European Coatings Journal*. (01), 52–56, (2018)
[2] Osterhold, M. Patterns of roughness. *European Coatings Journal*. (07/08), 44–48, (2016)
[3] Osterhold, M. Improving rheology measurement. *European Coatings Journal*. (11), 48–52, (2016)

The concept of polymer brushes for the next generation of multifunctional coatings

Alexander S. Münch[1], Susanne Höhne[1] and Petra Uhlmann[1]

Leibniz-Institut für Polymerforschung Dresden e.V., Hohe Straße 6, 01069 Dresden, Germany, e-mail: uhlmannp@ipfdd.de

Due to the specific physicochemical properties of responsive polymer brushes, which have been intensively investigated regarding fundamental aspects in the last decades, this form of surface modification is also promising for various applications including wettability control, sensing, controlled bio adhesion, catalysis, and biomedical applications. Although extensive fundamental investigations lead to a multitude of scientific publications, polymer brushes are rarely transferred to commercial use. The reasons for that are neither their excellent properties, nor their modularity or the understanding of responsiveness but rather the transfer of preparation from small model surfaces, used in the laboratory, to technically relevant substrates and coating processes. This includes material scalability, short coating times at industry-relevant low temperatures in a single process step, and the use of typical technical coating processes. Often, scientific methods for generating polymer brush coatings do not meet such requirements.

We present a general way showing how it is possible to prepare functional polymer layers under process-relevant conditions for technical purposes. In this context, the classical concept of polymer brushes has to be altered. It will be shown that it is necessary to deviate from the typical brush architecture without losing the specific properties, such as switching by external triggers. Using different exampled of functionalized *co*-polymers, multifunctional coating with anti-fouling, easy-to-clean, anti-fog and anti-icing properties will be presented, which fulfills the requirements of technical coating processes [1,2,3,4]. It will be shown that in addition to application-relevant issues, a deep fundamental understanding of responsivity is crucial for the generation of the desired properties. This will be exemplarily demonstrated by the easy-to-clean functionality [5].

[1] Münch, A.S., Wölk, M., Malanin, M., Eichhorn, K.-J., Simon, F., Uhlmann, P., *J. Mater. Chem. B* 6 (5), 830 – 843 (2018)
[2] Münch, A.S., Stake, A., Lukasczyk, T., Stalling, A., Fritzsche, T., Stenzel, V., Uhlmann, P., *J. Coat. Technol. Res.* 16 (6), 1515 – 1525 (2019)
[3] Münch, A.S., Fritzsche, T., Haufe, H., Uhlmann, P., *J. Coat. Technol. Res.* 15 (4), 703 – 712 (2018)
[4] Höhne, S., Hoch, C., Böhm, C., Winkler, R., Bittrich, E., Uhlmann, P., *Langmuir* 36 (16), 4465 – 4476 (2020)
[5] Münch, A.S., Adam, S., Fritzsche, T., Uhlmann, P., *Adv. Mater. Interfaces* 7 (1), 1901422 (2019)

Acknowledgement: The authors gratefully acknowledge the funding from the Federal Ministry for Economic Affairs and Climate Action (BMWK) of Germany (AiF-IGF 18696 BR and AiF-IGF 18573 BG/1) and from the Federal Ministry of Education and Research (BMBF) of Germany (Grant 03SF0475B).

Exalt® water-based calcium aluminate cement suspension for fast drying in water-based coatings.

E.Bertin[1], E.Giraud[1], E.Pena-Amiel[1]
[1]S&T, Imerys, Toulouse, France, Emmanuel.bertin@imerys.com

Calcium aluminates are recognized formulation levers in the drymix mortar industry. When combined with calcium sulfate or Portland cement, these powdered hydraulic binders react with water and deliver key functions such as rapid drying and hardening; high mechanical strength; controlled dimensional stability and high-end aesthetics. Such performance can now be achieved in water-based "ready to use" products with the unique Exalt® technology. Exalt® is a patented concept of calcium aluminate stabilization in a water-based suspension. This extends the application of these materials and their benefits into pastes; adhesives; and paints and coatings, opening new uses and functions in these hybrid systems (organic – mineral based).

Exalt® water-based suspensions are safe and easy to use. They contain high active solids content (60% in weight), have low viscosities (500-4000 mPa.s), are non-aggressive (pH 5 to 7) and are highly stable. Exalt® containing formulations can then be easily "activated" by adding an alkali source to begin the curing process. Further modifications with other additives may be used to adjust the setting time.

As Exalt® is reactive in cold and humid conditions, one of the first identified benefits has been for architectural coatings applications. This paper reports on the use of Exalt® calcium aluminate suspension in water-based exterior renders and especially the benefits it brings for fast drying in difficult environmental conditions such as relatively low temperatures and high relative humidities. Other properties related to improved drying properties, such as early rain resistance, have also been evaluated. There is no standardised testing procedure to measure drying speed in such conditions, thus an "in house" method was implemented combining drying in a climate-controlled oven and immersion tests. The approach of the study was to evaluate the impact of various key formulation parameters on such hybrid-based systems, including the activation alkali source, activation pH and use of accelerators.

Anti-corrosive properties of organofunctional alkoxysilane coatings

K.Szubert, *H. Maciejewski*

Faculty of Chemistry, Adam Mickiewicz University in Poznan,
Uniwersytetu Poznańskiego 8, 61-614 Poznan, (Poland) (country),
e-mail: karolszu@amu.edu.pl

Metals, such as iron, aluminium, copper and magnesium and their alloys are used in a myriad of structural, marine, aircraft applications and cultural heritage, etc. While these metals are useful because of their physical characteristics, such as stiffness and high strength to weight ratios, they are highly susceptible to corrosion in aggressive environments. Corrosion is always the major reason of energy and material loss. The need for toxic chromate replacement in metal-finishing industries has prompted intensive search for new technologies in resent year. For example, trialkoxysilanes start to displace chromates(VI) for the production of metal's protective layers. New technologies are compatible with the idea of using environmentally friendly compounds, as they are free of chromate pigments, chromium, or volatile organic compounds [1,2].
The use of silanes for metal protection against corrosion had been foreseen a long time ago [3], but intensive research in this field has only been carried out within the last few years. The role of silanes in protection against corrosion is associated with formation of a protective coating layer which exhibits a barrier effect against the corrosive factors and is capable of inhibiting the corrosion processes. The silanes bound to the metal surface form a protective layer with organic groups oriented towards the outside of the metal surface; mainly through the hydrophobic properties formed layer protects metal surface from contact with the electrolyte (water), and thus prevent the creation of corrosion cells [4].
In this communication we present a new trialkoxysilanes derivatives synthesised from fatty acids salts and their application in anticorrosion coatings

References:

[1] van Ooij, W.J., Zhu, D., Prasas, G., Jayaseelan, S., Fu, Y., Taredesai, N,, Silane based chromate replacements for corrosion control, paint adhesion, and rubber bonding, *Surf. Eng.,* 16, 386-396 (2000)

[2] Materne, T., de Buyl, F., Witucki,G.L., *Organosilane Technology in Coating Applications: Review and Perspectives*, (2012) Dow Corning Corporation

[3] Jones, D. A., *Principles and prevention of corrosion* (1996) Pearson

[4] van Ooij, W.J., Zhu D., Corrosion protection properties of organofunctional silanes - An overview, *Tsinghua Science and Technology*10(6) 639-664, (2005)

Solar control coatings for heat harvesting of building façades

Bart Erich (TNO), Corry de Keizer (TNO-SEAC), Anthonie Stuiver (AkzoNobel), Pascale Rigouts (AkzoNobel), Sanne van Leeuwen (TNO), Jonathan van den Ham (TNO), Roland Valckenborg (TNO-SEAC), Nicole Meulendijks (TNO), Gerrit-Jan van Riessen (Emergo)

The Netherlands Organisation of Applied Scientific Research, High Tech Campus 25, 5656 AE Eindhoven, The Netherlands

Within the framework of the Horizon 2020 project ENVISION we have developed a new invisible façade integrated collector panel with a unique colored, solar harvesting coating. This coating system and façade panel is developed with paint company AkzoNobel, together with prefab element producer Emergo and The Netherlands Organization of Applied Research (TNO).

The potential of thermal collectors integrated in the façade in the Netherlands is surprisingly high with an annual energy solar gain of 2.5 – 3 GJ/m^2 in a south oriented façade and still a 1 GJ/m^2 in a north facing façade. In a novel well insulated row dwelling in the Netherlands, 7 GJ is needed for generating tap water and 7 GJ is needed for heating. Taking an efficiency between 40% and 95% when the façade collector panels are at ambient temperature. This means between 5 and 10 m^2 is sufficient to generate the required energy for heating and tap water production, which is roughly the façade area available per apartment. In combination with heat pumps and roof equipped with PV the building will be easily energy positive. The only challenge is the storage.

Within the framework of the ENVISION-project (www.energy-vision.eu), the performance of these façade collectors will be tested at the Solar Energy Application Centre (SEAC) of TNO. The performance of these panels is measured for different coating solutions. Laboratory tests have shown that even for the white façade collector up till 40% total solar absorption can be achieved, by our nano-technology modified coating system.

This means that we can include the technology seamlessly in all buildings with different colours. It will be a direct competitor for air heat pumps and water/water heat pumps with ground loops.

After testing the solution in SEAC, the technologies will be demonstrated in full demonstrations in Italy and the Netherlands. In Italy, the demonstration focuses at connecting the harvesting panels to a district heating network of the University of Genoa. End of 2019 in the Netherlands a first demonstration project has been realised in the city of Almere. First results on the experience of this demonstration will be available at the moment of the conference.

In a presentation at the conference, the status will be shown, including the performance of the building integrated panels, as well as the planned demo's including its potential for buildings in the future.

Mechanisms of protection by coatings with a reduced content of zinc pigments

Ewa Langer[1], Małgorzata Zubielewicz[1], Agnieszka Królikowska[2], Leszek Komorowski[2], Matthias Wanner[3]

[1]The Łukasiewicz Research Network – Institute for Engineering of Polymer Materials and Dyes, Toruń, Poland, e.langer@impib.pl, m.zubielewicz@impib.pl
[2]Road and Bridge Research Institute, Warsaw, Poland, akrolikowska@ibdim.edu.pl, lkomorowski@ibdim.edu.pl
[3]Fraunhofer-Institut für Produktionstechnik und Automatisierung, Stuttgart, Germany, matthias.wanner@ipa.fraunhofer.de

In zinc-rich primers Zn acts in the initial stage as sacrificial anode (cathodic protection). During lifetime Zn is transformed into its corrosion products, which seal the coating's pores (barrier protection). Such coatings are typically very highly pigmented which could cause the deterioration of cohesion and/or adhesion as well as mechanical properties. Therefore – and because Zn is known to be harmful to the environment – research works to develop Zn primers with reduced Zn content but good corrosion protection were undertaken. The formulations of different Zn pigments in epoxy matric were developed with concentration of zinc particles reduced to about 50 mass% in dry coating. Following Zn pigments were used: mixtures of zinc pigments with different particle size and particle shape, zinc dust with organic surface treatment, ZnMg and ZnSn pigments as well as Zn dust with an addition of graphene or carbon nanotubes. Mechanism of protection by coatings with a reduced content of zinc pigments have been studied in comparison with commercial zinc rich primers – epoxy and ethylsilicate – by the following methods: Electrochemical Impedance Spectroscopy (EIS), Scanning Kelvin Probe (SKP), EIS / Water Uptake Reversibility (WUR), Salt Spray Test (SST) and the exposure in the natural marine environment, in splash salt water zone. It was found that cathodic protection was shown by primers with mixture of Zn flakes and extenders as well as Zn dust and graphene. Primers pigmented with ZnSn or Zn dust with carbon nanotubes (CNT) keep their corrosion potential in the transition phase. Other primers show only barrier protection. Among the zinc-rich commercial primers, cathodic protection was observed for ethyl silicate primers and one of the epoxy primers – an innovative paint with a high content of activated zinc. Coatings characterized by corrosion potential in the range of cathodic protection or in the transition phase have very good protective properties in the corrosive environment. However, good anticorrosive properties were found also in the case of primers which do not show any cathodic protection, such as the primer pigmented with the mixture of Zn particles of different size and the primer containing Zn dust with chemical surface treatment. It shows that good anticorrosive properties can be obtained even if cathodic protection is not observed.

Acknowledgments

The research was carried out under the project CORNET/22/1/2016 "New generation of zinc primers with improved anticorrosion, application and ecological properties (ZINCPOWER)" funded by the National Centre for Research and Development in Poland and the AiF within the programme for the funding of the industrial research (IGF) in Germany.

Novel polyetherimide (PEI) antifriction coatings with chemical bonded PTFE as lubricant

M. Gedan-Smolka[1], M. Malanin[1], H. Großmann[2], T. Schmidt[1], A. Garbe[1]

[1]Leibniz-Institut für Polymerforschung Dresden (IPF), Hohe Str. 6, 01069 Dresden, Germany, mgedan@ipfdd.de
[2]Anton Paar GmbH, Hellmuth-Hirth-Str. 6, 73760 Ostfildern-Scharnhausen, Germany

A low friction coefficient causes the use of polytetrafluoroethylene (PTFE) as solid lubricant in antifriction coatings as well as in tribological bulk materials. However, it is well known that PTFE is not compatible to almost all plastic materials. For that reason a homogeneous distribution and breaking down of PTFE particles in a polyetherimide (PEI) matrix to generate anti-friction coatings is not successful without adding of any dispersion stabilizer so far.

The objective of our research was to use γ-irradiated and, for comparison, non-modified PTFE-micro powder to generate PEI-PTFE antifriction coatings and to compare several procedures to get an improved breaking down behavior and homogeneous distribution of the PTFE particles in the matrix material.

Two different approaches were studied:

- Reactive extrusion of γ-modified PTFE with PEI followed by a high power dispersion procedure
- High power dispersion of γ-modified/ non modified PTFE

It was already found in the past that a defined high-energy irradiation of PTFE powder results in a generation of reactive functional groups or persistent radicals [1]. These species are able to form chemical bonds to other polymer matrices, e.g. polyamideimide [2,3]. The high shear rate in the extrusion procedure enables a higher bonding degree in comparison to a high power mixing of both components during dispersion. The chemical bonding was proved by FTIR-spectroscopy and SEM micrographs. The thus obtained PEI-PTFE-cb pellets were subsequently used for the formulation of solvent-based dispersions. The dispersions were characterized with regard to rheological properties, dispersion stability and particle size distribution. In addition, they were compared to state of the art physical mixtures of unmodified PTFE in a PEI matrix.

It was demonstrated that chemical bonding between PTFE and PEI results in very stable dispersions without adding of any dispersion agent at all. Furthermore, the resulting particle size is much lower and the particle size distribution is more homogeneous than in physical mixtures of non-modified PTFE-micro powder. The novel, chemical bonded, coating material was applied to steel substrates afterwards. Highly flexible, smooth and glossy films with a homogeneous PTFE-particle size distribution and improved tribological properties were kept.

References:

[1] K. Lunkwitz, U. Lappan, U. Scheler, "Modification of perfluorinated polymers by high-energy irradiation", Journal of Fluorine Chemistry, vol. 125, pp. 863-873 (2004)

[2] M. Gedan-Smolka, et al., "Influence of chemical bonding of PTFE lubricant to a polyamideimide (PAI) matrix in antifriction coatings," in World Tribology Congress, pp. 1–4 (2017)

[3] Gedan-Smolka, M., A. Marschner, D. Lehmann, R. Franke, M. Klemm, E. Leidich, B. Prase: „Formulation and Characterization of PAI-PTFE-cg Antifriction Coatings", Key Engineering Materials, ISSN:1662-9809, Vol. 721, pp 362-367 (2016)

Iodonium Salts with Tunable Efficient Chromophores as One-Component Cationic Photoinitiators for 3D Printing

F. Petko[1,2], M. Galek[2], E. Hola[1], J. Ortyl[1,2,3]

[1]*Faculty of Chemical Engineering and Technology, Cracow University of Technology, Warszawska 24, 31-155 Cracow, Poland*
[2]*Photo HiTech Ltd., Life Science Park, Bobrzyńskiego 14, 30 348 Cracow, Poland*
[3]*Photo4Chem Ltd., Lea 114/416A, 416B, 30-133, Cracow, Poland*
e-mail: filip.petko@doktorant.pk.edu.pl

Materials based on the cationic photopolymerization of monomers, such as vinyl ethers, epoxides, oxetanes, and glycidyls, exhibit favorable properties such as the fast cure, low shrinkage, good adhesion, good electrical properties, and no oxygen inhibition [1]. That is the reason for the growing interest of cationic photopolymerization in many fields of industry, especially in 3D printing [2].

One of the most active cationic photoinitiators are diaryliodonium salts. However, their absorption properties are negligibly low above 300 nm what makes them inefficient or inactive under irradiation emitted by *Light Emitted Diodes* (LEDs). Some modification of diaryliodonium salts structure have been made but with little effect on absorption properties [3]. The use of more complex chromophores had better results. One of such chromophores was 7-methoxycoumarin which red-shifted iodonium salts absorption bands above 350nm. This type of iodonium photoinitiators is characterized by efficient photolysis under LEDs irradiation and high quantum yield of photoacid generation. Therefore, initiators of this type are suitable for photoinitiating the cationic polymerization of monomers such as epoxides, vinyl ethers, oxetanes and glycidyl ethers.

Here, we present a new group of photoinitiators based on iodonium salts equipped with efficient chromophores [4]. New type of chromophore allows iodonium salts to absorb light in UV-A and visible light region of spectrum and makes them active toward epoxides or vinyl ethers under irradiation of LEDs emitting light at 365nm and 405nm. Such properties, suitable for applications in 3D printing are ensured by push-pull effect in D-π-A structure of chromophores. Presented scaffold of chromophores can be easily modified and absorption properties of iodonium salts equipped with such chromophores are easily tunable.

Acknowledgments

The work resulted from the implementation of the research project No. 2021/41/N/ST5/04454 funded by the National Science Center"

References

[1] Z. Chen, Y. Zhang, B. Chisholm, D. Webster, A humidity blocker approach to overcoming the humidity interference with cationic photopolymerization. *J. Polym. Sci. A – Polym. Chem.*, 46, 4344-4351 (2008)
[2] S. C. Ligon, R. Liska, J. Stampfl, M. Gurr, R. Mülhaupt, *Chem. Rev.*, 117, 10212-10290 (2017)
[3] M. Topa, E. Hola, M. Galek, F. Petko, M. Pilch, R. Popielarz, F. Morlet-Savary, B. Graff, J. Lalevée, J. Ortyl, One-component cationic photoinitiators based on coumarin scaffold iodonium salts as highly sensitive photoacid generators for 3D printing IPN photopolymers under visible LED sources. *Polym. Chem.*, 11, 5261-5278 (2020)
[4] F. Petko, M. Galek, E. Hola, R. Popielarz, J. Ortyl, One-Component Cationic Photoinitiators from Tunable Benzylidene Scaffolds for 3D Printing Applications. *Macromolecules*, 54 (15), 7070–7087 (2021)

Renewable 100% bio-based 1,3 propanediol (PDO) as building block in polyurethane dispersions for polyester and polyether polyols.

Patrick Van Waes, Sean L. Gahan

DuPont Tate&Lyle BioProducts, Loudon Tennessee United States of America

The use of alternative bio-based sustainable and renewable building blocks in the coatings industry is more than ever a strategic choice paint producers and resin producers are making today. 100% bio-based building blocks replacing fossil fuel alternatives commonly used are challenged by the need of equal to better performance and being price competitive at the same time.

This paper will elaborate about the usage of an industrial corn based 100% bio-based 1,3 propanediol (PDO) in polyurethane dispersions as a building component of polyester and polyether polyols and also as a chain extender. Using PDO will result in a substantial bio component in waterborne PU coatings formulations with similar quality properties and lower environmental footprint and dependency on fossil fuel.

PUDs based on polyols derived from 1,3-propanediol (PDO) were evaluated as direct to metal (DTM) coatings.

Two component waterborne PUDs were produced from PDO based polyester polyols (PDO-Adipate 2000MW, PDO-Sebacate 2000MW, and PDO-Dodecanedioate (C12) 2000 (MW) and PDO derived polyether polyol (Velvetol H 2000 polyether polyol from Allessa) and applied onto five metal substrates (steel, stainless steel, aluminum, aluminum 2024 and aluminum 7075) to evaluate their performance and commercial viability coating applications. The impact of PUDs based on PDO polyols was evaluated on the film properties, adhesion strength, coating adhesion, corrosion of metal substrates, and solvent resistance. The performance of PUDs based on PDO polyols was also compared to reference PUDs based on conventional butanediol-adipate 2000MW polyester polyol, poly(oxytetramethylene) glycol 2000MW (PTMG 2000), and Polycarbonate diol 2000 MW.

The bio-based coatings prepared in this investigation provided useful physio-mechanical properties, chemical resistance, and water resistance compared to reference and commercial petroleum based two-component (2K-PUDs).

The results of this study show that PDO derived polyester polyols, while sustainable and from 100% bio-based industrial corn feedstock origin, also provide similar performance in 2K PUDs that could benefit coatings direct to metal applications.

REFERENCES

1. H.X. Xiao, K.C.Frisch, ed.: Advances in Urethane Ionomers., Inc. Technomic Publishing Company Inc., Lancaster, Pennsylvania, USA; 1995
2. Z. W.Wicks, Jr., D.A.Wicks, J.W. Rosthauser, Progress in Organic Coatings, Vol.44 (2002)
3. B.K.Kim, J.C Lee., J. Polymer. Sci., Polym. Chem. Ed, Vol. 34 (1996), 1095-1104.
4. R. Miller (DuPont Tate & Lyle Bio Products), R. Janssen and L. Theunissen (Reverdia)," Evaluating the Properties and Performance of Susterra 1,3-Propanediol and Biosuccinium Sustainable Succinic Acid in TPU Application", CPI Polyurethanes 2012 Technical Conference, September 2012, Atlanta
5. S.J. Hurff and M. Shen (DuPont Tate and Lyle Bio Products), I. Sendijarevic and A. Sendijarevic (Troy Polymers), "Advances in the Use of Bio-based Components in Aqueous Polyurethane Dispersions for Coating and Adhesive Applications", CPI Polyurethanes 2018 Technical Conference
6. S. L. Gahan and M. Shen (DuPont Tate and Lyle Bio Products), I. Sendijarevic and A. Sendijarevic (Troy Polymers), "Evaluation of 1,3-Propanediol Based Polyester and Polyether Polyols in Polyurethane Water Dispersions (PUDs) for Textile Coating Applications", CPI Polyurethanes 2019

Sustainable, Bio-based Polyester Resins for Coatings

R. Wollrab, *N. Čuk, S. Zabret, M. Ocepek*

Helios, Domžale, (Slovenia), radmila.wollrab@resinshelios.com

Driven by the regulations and consumers, an ever-increasing push towards sustainability and a reduced carbon footprint has inevitably influenced the polyester resins industry. The meaning of the term “environmentally friendly” in relationship to coatings and consequently resins has evolved through the years. Initially, the goal was pollution prevention or reduction. This meant reducing solvents in formulations, producing high solids coatings, waterborne and 100% non-volatile coatings such as powder coatings. Nowadays every stage from the production to the end-of-life is considered – from raw material manufacturing, formulations to application, recycling, and disposal. Helios Resins is investing in the research and development of new bio-based resins produced from bio-monomers obtained from a huge range of biological carbon sources. In each case, the origin is biomass that is broken down using adapted chemical processes in a bio-refinery to produce biochemicals that lead to bio-monomers. Our projects range from replacing petro-based terephthalic acid with 2,5-furandicarboxylic acid (FDCA) or recycled terephthalic acid in bio-polyesters for coil to replacing petrol-based adipic acid with bio-based succinic or sebacic acid in bio-polyesters used in 2K PUR coating formulations. Other activities include the use of bio-isosorbide and bio-based 1,3-propanediol, 1,4-butanediol, 1,5-pentanediol. Moreover, we are already using monomers that are partially or entirely bio-based in our conventional products, mostly polyesters. Such examples are neopentyl glycol, trimethylolpropane and pentaerythritol. In this work we give an overview of the drop-in and novel bio-based, sustainable resins and the ISCC plus certification that allows us to determine and communicate bio-based content to our customers.

Key words: sustainability, polyesters, bio-based, coatings

Improved Anti-Fouling Performance in Cuprous Oxide based coatings through incorporation of modified fumed silica

M. Heuer, J. Foth, K. Roland

Evonik Operations GmbH, Essen, Germany,
e-mail: marco.heuer@evonik.com

More than 10 years ago structure modified synthetic silica technology was introduced for improving scratch resistance and mechanical strength in coatings. New developments now offer environmentally sustainable improvement to anti-fouling performance in marine environments.
Due to biofouling CO_2 output is tremendous. To avoid biofouling biocidal copper oxide based anti-fouling coatings are mostly used, but they are more and more under criticism. Typically, very high loadings of copper oxide biocide are needed for full performance, sometimes higher than 50 weight-%. Incorporating a novel structure modified, nanostructured synthetic silica, the amount of copper oxide can be reduced dramatically, while the coating still offers full anti-fouling performance.
This paper addresses the effect of structure-modified, nanostructured silica in marine anti-fouling coatings. It gives results on the anti-fouling performance from field tests. Special focus will be drawn on the mode of action of the product.
Furthermore, this paper explains how to use structure modified fumed silica technology with cuprous oxide to improve the antifouling behavior and life-cycle performance of antifouling coatings.

Ionic Liquids Curing Agents as an ideal tool in development of composite and biocomposite materials technologies

Marcin Smiglak[1], Dawid Zielinski[1,2], Andrea Szpecht[1], Joanna Dzialkowska[2], Paulina Nadobna[1]

[1]Poznan Science and Technology Park, Poznan, Poland
[2]Adam Mickiewicz University, Poznan, Poland
Email: marcin.smiglak@ppnt.poznan.pl

Epoxy resins for over 70 years are used in many industries, and because of their unique properties, they are considered one of the most important groups of polymers. They have chemical resistance and good mechanical and thermal properties, and have found a number of practical applications, such as adhesives, coating materials, paints, composites, molding materials, microelectronic materials or matrixes for fiber reinforced composites [1,2].

Ionic Liquids (ILs) are becoming increasingly popular group of chemicals, applied in many branches of science and industry [3,4]. One of the interesting properties of ionic liquids is their potential ability to be used as hardeners for epoxy resins [5]. The IL in this case does not act on its own as a hardener, but becomes one when heated to a certain temperature.

Fundamental studies presented in this work are mainly focused on utilization of ionic liquids and their eutectic mixtures, as a new initiators of crosslinking of the epoxy resins, as well as on the investigation of the influence of the applied filler type on the physicochemical properties of the obtained composite material.

These studies are focused on the analysis of data from thermal and strength analyzes for many ionic liquids representing various groups of these compounds. In addition, both ionic liquids used directly as curing initiators, as well as in eutectic systems, liquefaction systems, solid systems and curing-modifying systems that affect the strength of the final materials will be analyzed. The research was carried out on a group of many ionic liquids in systems with commercial resins and epoxy bio-resins, and DSC tests and mechanical SEM tests were used for the analyzes.

This work was supported by the National Science Centre (Poland), project SONATA BIS no UMO-2017/26/E/ST8/01059.

References:

1. J. P. Pascault, R. J. Williams General concepts about epoxy polymers. in: J. P. Pascault, R.J. Williams (editors) Epoxy polymers. New materials and innovations, Wiley-VCH Verlag GmbH & Co, Weinheim; 2010, p. 1–12.
2. H. Q. Pham, M. J. Marks Epoxy resins. In: H. F. Mark (editor) Encyclopedia of polymer science and technology, vol 9, John Wiley & Sons, New York, 2004, 678–804.
3. C. Chiappe, D. Pieraccini Arkivoc 2002, 11, 249-255.
4. F. Endres ChemPhysChem 2002, 3, 144-154.
5. M. A. M. Rahmathullah, A. Jeyarajasingam, B. Merritt, M. VanLandingham, S. H. McKinight, G. R. Palmese Macromolecules 2009, 42, 3219-3221.

VOC-Toolbox – More Freedom for Coatings Formulators

Marco Heuer, Andreas Stuettgen, Sascha Herrwerth

Evonik Resource Efficiency GmbH, Essen, Germany,
e-mail: marco.heuer@evonik.com

Worldwide the demand for more eco-friendly coatings is still increasing after US-American and European initiatives for reducing the VOC emissions several years ago also China decided in 2017 to join this way.
A VOC-toolbox of new developed binder technologies, co-binders, hardeners, specialty fillers and additives based on various chemistries like silicone-hybrid-resins as well as polyester based resins, new structure of polysiloxane-additives, special precipitated silicas and new blend of 2,4,7,9-Tetramethyl-5-decin-4,7-diol offer a unique combination of low viscosity and high durability.
Next to these properties also improvements like enhanced adhesion promotion, higher weather stability as well as excellent color stability are achieved by using these raw materials. It is possible to create high-solid and ultra-high-solid coatings materials easily or even non-VOC-formulations.
Restrictions in Scandinavia about a limited usage of isocyanate curing products in the coatings industry narrowed the portfolio of suitable resins. Also BPA-based materials are under general suspect nowadays in Europe and for example banned in France since 2018 in food containers. Special silicone-hybrid-resin technologies with a non-isocyanat-curing mechanism based on hydrolysation-condensation reaction allows a durable polymeric network for protective coatings. In conjunction with aliphatic alpha-omega-trialkoxy-urethane resins it is also possible to create different degrees of hardness without losing any of the excellent properties like weather or chemical resistance.
A new technology based on a special designed hardner for Methylphenylsilicone resins allows the formulator to design formulations with excellent adhesion on various substrates, which achieves corrosion protection combined with heat resistance up to 650°C in a direct-to-metal formulation combined with outstanding UV-resistance.
Solvent-free adhesion resins and dramatically VOC-reduced grinding resins offer also savings regarding the emission of solvents during application and flash-off time, while keeping viscosity on a low level.
The new developed and patented 2,4,7,9-Tetramethyl-5-decin-4,7-diol based surfactant without any VOC-release completes the VOC-Toolbox next to 100% additives based on polysiloxane technology range.
So more freedom and options for the formulator is given without any invest into new production or application equipment.
By using ingredients of the VOC-Toolbox it is possible to achieve ready-to-apply coatings with 90% solid content.
These different technologies allow the fulfillment of common and future legal restrictions of different coatings materials.

CARELESSNES IN MORPHOLOGY DESIGN CAN AFFECT ORGANIC COATINGS PERFORMANCE

Agnieszka Królikowska[1], Leszek Komorowsk [1]i, Pier Luigi Bonora[2]
[1]Road and Bridge Research Institute, Warsaw, Poland, akrolikowska@ibdim.edu.pl; lkomorowski@ibdim.edu.pl; *[2]AITIVA, Piacenza, Italy,* bonorapierluigi@gmail.com

Abstract

Protecting coatings for structural steel are formulated and applied in billions of square meters to the aim of avoiding or at least delaying both aesthetic and structural degradation, providing a predictable long service life to bridges, plants, ships, cars. Inside the most countries of European Community paints are not classified as „building materials", hence not subjected to control by certified laboratories, but left to the application of „International Standard procedures" agreed between parties. Standards cover some specific properties and features and relate to "requirements" also agreed by the parties according to the expected durability in a classified aggressiveness of an environment. The Standards are mainly, if not exclusively, concerned with performances and related to such parameters as both number and thickness of layers, hardness, adhesion (expressed as the strength needed to remove the paint from the substrate), gloss, scrub resistance, corrosion resistance, porosity etc. The resin to pigment ratio, the surfactants, the quality of pigments and fillers (dimensions, shape, wettability, reactivity, homogeneity of distribution…) and their repeatability are not clearly required, defined and consequently checked during utilisation, where the applied paint might be different (by mistake or voluntary action) from the sample checked and accepted by an accredited laboratory. The instable and complex equilibrium among the very many chemically reactive and specifically useful components is at the basis of the quality of the final product, the paint. The above features should therefore be checked and included among possible causes of failures. The Anticorrosion Laboratory of IBDiM analysed widespread adhesion failures on two very huge structures checking many features of paint both composition and structure with both SEM and EDAX carrying out a comparison with a carefully formulated paint. In the first case, concerning partly immersed /partly exposed to marine atmosphere structures, the Laboratory collected several paint samples recovered after detachment from various locations of the structure. A comparison was possible of such samples with other ones obtained using paints of the same label either present in the Laboratory as reference (certified HD IM according to ISO12944 Standard) or bought directly on the shop.

A second case concerned a structure inserted in a polluted urban environment where extended delamination occurred between an epoxy intermediate and siloxane topcoat certified as ISO12944 class HD C5. Several detached samples were collected and examined. They were compared with samples of the same label taken from a construction where no failures were observed after twelve years of exposure in a HD C5 environment according to ISO12944.

The results showed morphological modifications as main responsible for the failures. where other causes were either absent or not dominating if paint features were not considered.

Key words: delamination, CPVC, pigments, fillers, extenders, topography

1K Crosslinking System for Polyurethane Dispersions

Wolfgang Paulus

BASF SE, Ludwigshafen, (Germany), e-mail: wolfgang.f.paulus@basf.com

Polyurethane dispersions (PUD) are widely used in the areas of industrial coatings and adhesives. They exhibit superior mechanical properties and good film forming with no or very little film forming agents. The production of the PUD is done in a 2-step process:

1. A prepolymer synthesis of the polyurethane is done either in substance or in solvents like acetone or NMP to a certain molecular weight.

2. Dispersing of the polyurethane in water. To achieve this, it is necessary that the PU backbone contains hydrophilic groups which stabilize the PU particles in water.

This process also shows the limitations of the PUD technology. In the first step, only limited molecular weights of the polyurethane are possible due to viscosity limitations. To obtain superior mechanical and chemical resistance properties in thin films it is necessary to increase molecular weights. This can partially be achieved by reacting remaining isocyanate groups with diamines after the dispersion step. But still the viscosity of the reaction mixture and the resulting dispersion is increasing, therefore it is not a generally used reaction pathway.

The alternative solution is to make the increase in molecular weight or even crosslinking within the thin film after film preparation. There are several commercial methods known to achieve this. Formulating the PUD with a water dispersable polyisocyanate leads to crosslinked films with superior film properties. But this is a 2-component application with limited pot life. The second approach is to incorporate acrylic moieties in the PU backbone, which can be polymerized after film formation with the UV curing technology. This is a stable 1K system but one needs UV- or EB light sources for curing.

A stable PUD with a potlife of 1 year which cures in thin films at ambient temperatures is a target, which is not realized today. In the field of polyacrylate emulsions, such a crosslinking system is already known for many years. Polymers containing carbonyl moieties (e.g. use of diacetone acrylamide (DAAM) in the monomer feed) can react with adipic acid dihydrazide (ADDH) via Schiff-base formation and thus crosslinking a thin film after drying.

In this paper we answer the question, whether it is possible to use a Michael- Addition reaction of dihydrazides, such as ADDH, to an acrylic double bond to crosslink a thin PUD film at ambient temperatures. Details are given about the kinetics of the crosslinking reaction in the dispersion itself and the dried films. Some results are shown about the property improvements of the dried films.

Additionally, others dihydrazides as crosslinkers are briefly investigated.

Effect of photo-initiator on physical and water vapor transport properties of polyethylene glycol diacrylate (PEGDA) based UV-cured acrylates

E. Jalilian*, G. Van Assche*, H. Terryn‡

*Vrije Universiteit Brussel, Physical Chemistry and Polymer Science
‡Vrije Universiteit Brussel, Electrochemical and Surface Engineering
Brussels, Belgium, guy.van.assche@vub.be

Research around UV-cured coatings as protective carrier layers for metals has been growing over the past decade [1][2]. UV-induced polymerization offers less demanding conditions in terms of processing, application and curing, and delivers a coating with tuned properties. Examples of such advantages are rapid curing, low energy requirements, low temperature treatments and the use of solvent free formulations [3]. Basic UV acrylate formulations must include photo-initiators that will form reactive species upon irradiation with photons of specific wavelengths and initiate the polymerization reaction. Despite being used in relatively minute amounts, the concentration of the photo-initiator and its chemical structure strongly affect curing kinetics and final properties of the layer.

In this work, the focus has been on the effect of the photo-initiator concentration on water vapor transport properties. For correct interpretation, physical properties of the layers were studied as well. This research is part of the *PredictCor* project that focuses on the development of a knowledge and technology platform for the prediction of durability and lifetime of organic coated metals against environmental corrosion.

To obtain information on physical properties, differential scanning calorimetry (DSC), attenuated total reflection infra-red spectroscopy (ATR-IR), thermogravimetric analysis (TGA) and dynamic mechanical analysis (DMA) were used. Water vapor permeation rates were determined using microcalorimetry. The results show that with increasing concentrations of the photo-initiator, higher reaction conversions are achieved. However, the glass transition temperatures (T_g) are comparable. Formulations with low concentrations of photo-initiator contain unreacted acrylates, while formulations with higher concentrations contain low molecular weight volatile segments. Microcalorimetry shows high permeation rates for layers cured with low and high photo-initiator concentrations, while the range in between shows lower permeation rates.

[1] D. Czachor-Jadacka and B. Pilch-Pitera, "Progress in development of UV curable powder coatings," *Prog. Org. Coatings*, vol. 158, no. April, p. 106355, 2021

[2] L. Ni *et al.*, "Direct-to-metal UV-cured hybrid coating for the corrosion protection of aircraft aluminium alloy," *Corros. Sci.*, vol. 89, no. C, pp. 242–249, 2014,

[3] G. Moad, *Chapter 01346 - Radical Polymerization*, no. April 2015. Elsevier Ltd., 2016.

Paint properties for oversprayfree painting technologies based on drop on demand principles

O. Tiedje, T. Hess, R. Nothhelfer-Richter, Q. Ye

Fraunhofer Institute for Manufacturing Engineering and Automation IPA, Stuttgart, oliver.tiedje@ipa.fraunhofer.de

Paint losses, called overspray, generated by spray coating applications are produced by paint not reaching the target and small droplets that will not deposit on the work piece. This overspray leads to manifold drawbacks since more paint is needed than necessary, contamination of the atomizer and booth may produce enhanced cleaning efforts and the risk of defects. To prevent health risks by inhalation of small droplets, the spray booth has to be carefully ventilated, which is one of the most prominent energy consumer in a paint shop. Furthermore, no precise coating is possible by spray application, so multicolour design and selective functionality of the surface usually create the need of masking processes.
Actually there are several developments to enable sharp edged paint application without any overspray. Most of these ideas involve the generation of single droplets by a nozzle [1]. Though the several years' experience in characterization of paints for spray applications are not satisfactory to deal with the challenges of these new technologies.
In the contribution the main properties of a paint for a drop-on-demand (DoD) approach will be described. The investigations are based on high speed videos of such an application process, enhanced viscosity measurements as visco-elastic properties, high shear rheometry and extensional viscosity. The fundamental background will be given by fluid dynamical simulations [2] yielding the droplet generation mechanism, the relevant shear and elongation rates and the pressure, resolved in time and space in the nozzle and the feeding tubes.
It will be present as well results for model substances showing how specific paint properties like extensional viscosity will cause different droplet forms as well as practical examples with commercial paints. Focus will be drawn to the satellite free droplet generation as well on the good levelling properties to avoid a structure due the drop by drop application process.

[1] Hess, T., Overspray-free painting technology – from the lab to the painting object. 90th Anniversary Conference of the Japan Society of Colour Material, Oct 2017, Tokyo
[2] Ye, Q., Rhein, S., Hess, T., Niemeier W., Tiedje, O., Investigation on viscous droplet generation for selective coating processes. 10th International Conference on Multiphase Flow, ICMF 2019, Rio de Janeiro, Brazil, May 19 – 24, 2019

Increase of powder coating quality and material efficiency by the use of a new smart control process

Markus Cudazzo, Dr. Karlheinz Pulli, Cathleen Joachimi, Mario Angos

Fraunhofer IPA, Stuttgart, (Germany), Markus.Cudazzo@IPA.Fraunhofer.de

A digitally transformed, Industry 4.0 (I4.0) capable, solvent-free powder coating requires amongst other things the use of an intelligent control system, in order to ensure a stable powder cycle with little or no transfer-efficiency-drift, layer thickness-drift, and particle size-drift. Methods already had been developed at Fraunhofer IPA in the 1990s, with which the powder recovery could be predicted by powder spray tests carried out in a powder coating line, accompanied by a therefore developed calculation method. Transfer efficiencies and particle size distributions of fresh powder and powder deposited on the workpiece have to be determined at a sufficiently representative workpiece spectrum. It is impossible to carry out powder coating trials in the powder coating facility during ongoing operation and the method is therefore associated with production losses. Accompanying experiments in a pilot plant and simulations at computer workstations are additionally necessary to receive a good overall-picture of the powder coating process, but they are also costly and time-consuming. They include, for example, simulation of powder spray-patterns in order to improve the overlapping, with the aim of achieving the highest possible layer thickness uniformity, up to quite scientific accompanying laser-Doppler measurements. Because the investigations described above are complex and most powder coaters are SMEs, process adjustments in the companies are almost empirical.

Coming back to the powder-cycle simulation, an important building block in terms of I4.0 capability, after a certain number of cycles, this method shows asymptotically the state of equilibrium between deposited and recovered powder. This behaviour emerges, when using the filter recovery, which is chosen for applications with a small number of colour changes to be carried out. When using cyclone recovery for factory processes with many colour changes, the recovery process is a bit more stable, but from the powder-coating processing companies a loss of (fine) powder material is deliberately accepted. With an adaptive technique, the intention is to increase the powder coating quality and material efficiency regarding the electrostatic powder spraying and to improve the controllability of the process.

To have a stable powder cycle from the beginning, the newly developed technique presented here, works with a combination of ejector and separator that separates particles based on gravity, centrifugal force, and settling velocity. Before feeding the powder from the hopper (mostly a fluidized bed) over a tube to the electrostatic powder spray gun, the powder is in a dense phase, which prevents the desired separation due to interaction forces, especially impact forces. To receive a thinned powder cloud with diminished interactions between the particles, additional air is injected into a small chamber. Using the settling velocity as a separation method, the coarser particles sink, while the finer particles are strongly suspended. With this design, the so-called vortex chamber-classifier leads to an intended non-selective but "easy to adjust" classification.

The results of the cognitive intelligent painting cell with the implemented vortex chamber-classifier show that the particle size distribution of the powder sprayed onto the workpiece corresponds to the fresh powder, so that the powder-cycle remains stable from the beginning.

References:

[1] Pulli, K.: "Optimierung der Pulverlackapplikation durch Anwendung experimenteller und numerischer Untersuchungsverfahren", Dissertation, Institut für industrielle Fertigung und Fabrikbetrieb IFF, Universität Stuttgart, ISBN: 3-936947-96-1, IPA-IAO Forschung und Praxis Nr. 443, Jost-Jetter Verlag, 2006, S. 15-17, S. 18-25, S. 26-37.

[2] Cudazzo, M.; Strohbeck, U.: „*Pistolenlose Pulverbeschichtungstechnologie für die platz- und prozesszeitminimierte Stückgutfertigung*", in: Abschlussbericht zum Verbundforschungsauftrag, Landesstiftung Baden-Württemberg, durchführende Stelle: Fraunhofer IPA, Stuttgart, 2009, Seite 4.

Colourful solar modules: Interference layer for low power loss

Dr. Udo Gumsheimer, Sebastian Barth,

Merck KGaA, Darmstadt, (Germany), udo.gumsheimer@merckgroup.com

Due to the exit of nuclear and fossil-fuel energy, solar technologies are getting more and more important. One of the barriers of implementing more solar into e.g. building skins or car bodies is the optical appearance of solar modules. By colouring solar modules, the acceptance at architects, house-builders, designers and car buyers can be increased, and the integration is much easier. Nevertheless, the efficiency of the modules may be heavily reduced by choosing the wrong colorants and thus the levelized cost of electricity would be increased. This would completely foil the efforts towards renewable energies, which is not acceptable.

The use of translucent interference pigments for the colouring of solar applications however can change the optical appearance of solar modules to a variety of attractive colours at low efficiency loss for cost competitive and optical appealing solar modules.

It will be shown that the integration of interference pigments respectively layers containing these substances in the production process of solar modules is easily manageable with a variety of possible technologies. Beside that it will be demonstrated that interesting colour shades are achievable while the solar modules stay efficient in producing electrical energy.

References: -

A novel method for coatings characterization at high temperature

Yassine Nagazi, Roland Ramsch, Giovanni Brambilla, Gérard Meunier
Formulaction, 3-5 Rue Paule Raymondis - 31200 Toulouse- France

Corresponding author: yassine.nagazi@formulaction.com

Characterizing the microscopic dynamics of coatings and polymer films is a very important step in every new development or quality control, especially at high temperature.

We use dynamic light scattering in the highly multiple scattering regime (Diffusing Wave Spectroscopy, DWS) to investigate the microscopic dynamics evolution of different coatings materials at constant or increasing temperature (RT-300°C) and to detect the phase transition temperature for different films with a very easy sampling protocol and a very high sensitivity for a large panoply of samples.

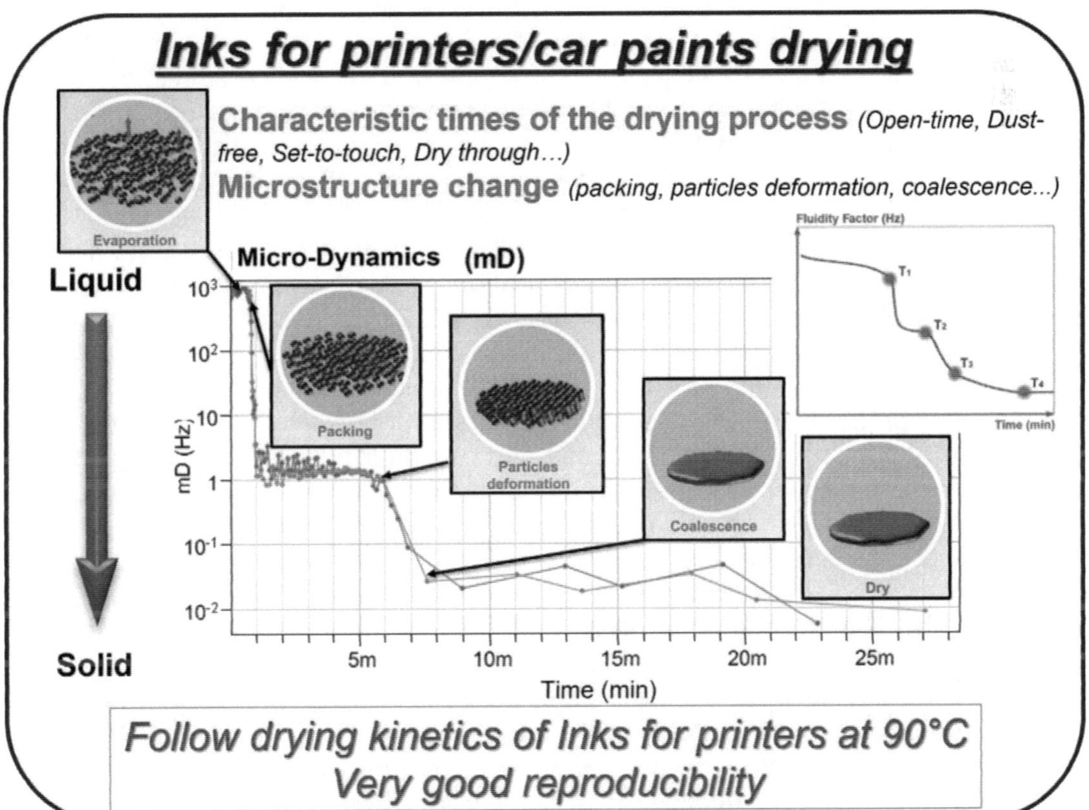

We propose a new, non-invasive, very sensitive and cost-effective method. Several examples show the influence of the temperature on the drying, the evaporation before curing, the curing and the phase transition, which will allow the optimization of materials development providing new insights.

Modern building materials for the removal of air pollutants

A. Rabajczyk, M. Zielecka
Research Centre for Fire Protection (CNBOP-PIB), Józefów, Poland, e-mail: arabajczyk@cnbop.pl

Air quality is one of the most important problems of the modern world, as it determines human health and changes occurring in other elements of nature, including climate change. For this reason, actions are taken to reduce the amount of harmful substances in the air. One such action is the use of building materials with special properties such as self-cleaning, self-disinfectant and/or photocatalysis.

The most significant recently developed building materials with selfcleaning surfaces, in particular biomimetic superhydrofobic or superhydrophilic surfaces as well as photocatalytic ones decomposing pollutants will be presented and discussed.

Modified building materials are increasingly used to make roads, tunnels, pavements, buildings. Depending on the demand, various additives are used, including nano and macro photocatalysts based on metals such as Ti, W, Ag, Au, coating materials with designed surface properties, based on polymers, ceramics, metals or hybrid composite systems, through replication of natural systems such as lotus leaf, rice leaf, butterfly wings. The use of rich chemistry of silicon compounds and reactive polysiloxanes allows the use of sol-gel processes. However, one of the most common substances is TiO_2, which, depending on the material, exhibits photocatalytic and self-cleaning properties. Combination with other substances, such as SiO_2 or Ag, allows obtaining material with photocatalytic properties and strengthening self-cleaning properties or obtaining a self-disinfection effect. TiO_2 modification with using non-metals like carbon or nitrogen, allows obtaining material with antibacterial and self-cleaning properties.

Thanks to photocatalysis processes, they contribute to reducing the degree of air pollution with compounds hazardous to human health, such as volatile of organic and inorganic compounds, such as VOCs or NO_x, or microorganisms. Combined with the self-cleaning process, the modified building materials give the opportunity to extend the lifetime of the photocatalysis process and extend the life of buildings. However, it should be taken into account that the air contains not only gaseous compounds, but also dusts, whose presence significantly affects the efficiency of photocatalysis and self-cleaning processes. In the case of sorption on the surface of dusts, active sites cease to fulfill their role and the reduction of pollution drops practically to zero. Therefore, the effectiveness of a given material should be verified in real conditions. Laboratory conditions do not provide the possibility of creating an air sample that reflects the composition of real air in different places (e.g. cities, industrialized areas, villages), the impact of temperature on air quality or changes in building materials that have a significant impact on the efficiency of photocatalysis and self-cleaning processes. It is therefore necessary to carry out further work on the modification of building materials in such a way that it is possible to remove contaminants at a constant level in various environmental conditions.

Innovative Coating Products with Graphene

B. Münzing[1],*

1 Sales Director, The Sixth Element (Changzhou) Materials Technology Co, Ltd, Plant 8, 9 West Tai Lake Avenue, Wujin Economic Development Zone, Changzhou, China,

**corresponding author: bernhard.muenzing@thesixthelement.com.cn*

Graphene: An introduction

Graphene, fully characterized since 2004 at the University of Manchester, is strictly speaking the atomic layer of graphite. Without any defects, graphene is highly electrical and thermal conductive and mechanically very strong. It is a carbon material, chemically inert, can be – with some efforts – functionalized and absorb other atoms or molecules due to its very large surface area of 2630 m^2/g. Shortly after 2004, companies started to develop different production processes of graphene like chemical vapour disposition (CVD) on substrates and routes starting from graphite using different oxidation-reduction processes. All these processes generate different kind of graphene products and when used in bulk applications like coating they normally are composed of more than one layer – in average 4 to 10 layers. While thickness is nanoscale, lateral size is µm-scale. To address health, safety and environmental questions of graphene, the industry is engaged in REACH registration and other regulatory activities.

Corrosion protection with graphene

Being chemically inert and conductive, graphene is a candidate for enhancing corrosion protection of coatings. Focusing on reducing the zinc-content of 2k epoxy zinc rich corrosion protection primers, The Sixth Element together with Chinese institutes developed a graphene type suitable for this application. Addition of 1 % of this graphene type to dry film is sufficient to enhance corrosion protection significantly. In extended testing, more than 3000 hours have been achieved in standard salt spray testing. Graphene acts as a barrier against penetrating water and oxygen and as a synergist to support the corrosion protection function of zinc due to graphene's electrical conductivity. Since 2016 such a system is commercially available first applied on the steel construction of an offshore wind energy plant in China.

In none-zinc based system it could also be shown, that addition of graphene enhances corrosion protection, mainly due to its barrier effect. Customers using already graphene confirmed the outstanding properties. Work continues to further elaborate the full potential of graphene as corrosion protection additive in high demanding environments.

Heat dissipation with graphene

Dissipating heat from complex structures is a challenge. In most cases traditional heat dissipation systems like copper or aluminium foils do not work. Alternatively, coatings or adhesives are used, filled with e.g. metal or ceramic particles (copper, silver, boron nitride). To achieve high heat dissipation properties significant amount of these particles must be added.

Based on the successful development of highly thermal conductive graphene heat dissipation film with thermal conductivities higher than 1000 W/mK, The Sixth Element has developed a suggested formulation using 12 % graphene, leading to highly thermal conductive coating systems.

Appraisal of Mathematical Modeling of Alkyd Resin Polycondensation Reactor

[1*]Chigozie Francolins Uzoh.

[1]Department of Chemical Engineering, Nnamdi Azikiwe University, Awka, Nigeria

Faculty of Engineering, Nnamdi Azikiwe University, Awka.

*Corresponding author: E-mail: cf.uzoh@unizik.edu.ng

Phone: +2348033762029.

Abstract

Alkyd resins are polyester polymers of fatty acids which, when spread over a surface in a thin coat, form a solid, cohesive, and adherent film. They are major component of most coating materials. Alkyd resin polycondensation process has been subjected to extensive theoretical, mathematical, simulation and experimental studies because of inadequate kinetic models. The analysis of alkyd resin polycondensation process is outstandingly essential for the design and development of new resins suitable for a broad range of application. However, experimental approaches are not only time consuming, but they are also burdensome, clumsy and costly. Researchers have, therefore, introduced mathematical modeling to control the alkyd reactor polycondensation processes and to elucidate the fundamental understanding of the polymerization process dynamics to design new alkyd resin in a shorter time with comparatively lower cost. This paper reviews the latest development in the mathematical and simulation technologies that have been reported for the alkyd reactor polycondensation kinetics. The scope of this reviews include the modeling and simulation techniques used for alkyd resin polycondensation processes and the future challenges.

Keywords: polycondensation, mathematical modeling, alkyd resin, optimization, reaction engineering, kinetics.

Computational fluid dynamics of particle break-up in a new type of industrial batch homogenizers

K. Wojtas[1], R. Adamek[2], W. Orciuch[1], Ł. Makowski[1]

[1] Warsaw University of Technology, Faculty of Chemical and Process Engineering, Waryńskiego 1, 00-645 Warsaw, Poland, e-mail: krzysztof.wojtas@pw.edu.pl
[2] ICHEMAD-Profarb, Gliwice, Poland

The literature review [1-2] and industrial expectations indicate the necessity to develop new mixers to obtain suspensions of desired particle sizes. Industrial homogenizers are usually characterized by high energy consumption, which is mainly influenced by process time and system geometry. In order to develop a new solution it was decided to develop a new, extended theoretical model describing particle distribution in tank homogenizers. The model is based on computational fluid dynamics (CFD) supplemented with population balance. CFD is a method that uses numerical analysis to solve problems that involve fluid flows. Thus, it provides easier means to examine the influence of process parameters (impeller geometry and speed, concentration, temperature, etc.) on the course of the process. However, CFD alone is not sufficient enough for the purpose of accurate particle break-up predictions. Only supplemented with the population balance, which describes the evolution of a population of particles in the system, this method can be considered as a viable extended model.

This paper is focused on application of a new model describing particle break-up to designing the new plant batch homogenizers. The new proposed model allowed identification of favourable geometry of impeller, process conditions and assessment of process efficiency.

The studies were conducted using a typical test suspension (titanium oxide in water). Particle size distribution and volume-weighted averaged size were measured experimentally using Beckman&Coulter LS 13 320 laser diffraction particle size analyser. The shape coefficient of the particles, which is necessary to include in the population balance, was determined basing on scanning electron microscope (SEM) images. Samples for both measurements were taken at different stages of the homogenizing process and from different regions of the experimental system.

The experiments of the new homogenizer were carried out in a cylindrical tank on a laboratory scale (10 L) and the numerical simulations were performed on both laboratory and industrial scale (200 L). Obtained results showed that the course of the particle break-up process can be successfully influenced both by process conditions (volume of the region of effective dissipation rate, particle residence time) and system geometry (mixing head design).

This work was supported by the National Science Centre, Poland (Decision No. DEC-2019/03/X/ST8/01261).

References:

[1] L. Xie, C. D. Rielly, G. Özcan-Taşkin. Break-Up of Nanoparticle Agglomerates by Hydrodynamically Limited Processes. *J. Dispersion Sci. Technol.* 29 (4), 573 – 579, 2008

[2] J. Ramírez-Muñoz, G. Martínez-de-Jesús, A. Soria, A. Alonso, L. G. Torres. Assessment of the effective viscous dissipation for deagglomeration processes induced by a high shear impeller in a stirred tank. *Adv. Powder Technol.* 27, 1885 – 1897, 2016

Teaching and Research at the Niederrhein University

C. Schmitz

Niederrhein University of Applied Sciences, Krefeld, (Germany),
Institute for Coatings and Surface Chemistry (ILOC)
christian.schmitz@hsnr.de

The chemistry department of the Niederrhein University in Krefeld offers the specialization on the topic of coatings technology. The studies are classified in the studies of the Bachelor of Engineering starting with content on coating technology after two years gaining experience in the basics of chemistry followed by the Master of Engineering going more in the depth on the specialization. The students acquire knowledge in the fields of raw materials, formulations for typical coating applications, technical processes in the manufacturing or application, and testing methods for characterization of coating materials. The structure of the studies schedules theoretical backgrounds and practical aspects in project management promoting soft skills in presentation, personality development and experience in the coating industry during bachelor and master thesis. Internationalization of the studies are in progress and the launch of an international Master of Engineering is expected in the nearer future.

In the recent years a PhD program was established at the Niederrhein University with an increasing number of graduates especially in the chemistry department and the field of coatings technology. The PhD studies are executed in cooperation with universities mostly in the nearer surrounding supervised by the professors in Krefeld. Tasks are research in the fundamentals of the Applied Sciences usually linked to the R&D of the coating industry partners and teaching of undergraduates.

The research is allocated at the Institute of Coatings and Surface Chemistry with smaller industrial projects running over several months up to projects with financial support from the federation, federated state or the European Union. Featured topics are investigation of formulations, syntheses of new materials, analysis of corrosion processes, radiation curable coating systems, data structuring and analysis. This gives PhD and master students the possibility to work on relevant topics for future science operating on latest lab equipment.

The coatings laboratory is also supported by the infrastructure of the Institute for Surface Technology, which provides high-throughput automation and the digital lab. The goal of this endeavour is the establishment of new development processes based on data science and digitalization of daily routines.

Objective and Unbiased Fouling Resistance Rating using Image Analysis

Morten L. Pedersen[a,b,c], Claus Erik Weinell[a], Burak Ulusoy[a,b,c], Kim Dam-Johansen[a].

[a]CoaST Research Centre, Department of Chemical and Biochemical Engineering, Technical University of Denmark, DTU, Building 229, 2800 Kgs. Lyngby, Denmark.
[b]Sino-Danish Centre Education and Research, Beijing 100093, China;
[c]Sino-Danish College, University of Chinese Academy of Sciences, Beijing 100049, China;
mlyped@kt.dtu.dk

Keywords: Biofouling, Fouling control coatings, Fouling resistance rating, image analysis

Fouling control coatings (FCC) are used to prevent the settlement of marine biofouling, which is the undesired accumulation of microorganisms, macroalgae, and animals on submerged surfaces. Biofouling increases the friction across the ship hull and causes a loss in speed, maneuverability, and fuel efficiency [1]. Another disadvantage of the accumulation of biofouling is the global spread of nonindigenous species, which is considered one of the four greatest threats to the biodiversity in the ocean [2]. In the development of new FCC, static immersion of test panels is the most common method to evaluate the performance. The static immersion is combined with regular visual inspections to document the coating performance. The are several evaluation standards utilizing observations from the visual inspection to determine a fouling resistance rating (FRR) as an indication of the FCC performance. However, the visual inspection will unavoidably induce some degree of subjectivity to the result. In the article by Pedersen et al. [3], the subjectivity related to visual inspection was illustrated, and a standardized biofouling recognition model using image analysis was proposed to provide a more objective basis for the evaluation of FCC. A pixel classification model was successfully trained in the program ilastik to recognize and generate a simple segmentation of three categories of biofouling, namely microalgae, macroalgae, and animals. The model showed a clear distinction between the different biofouling categories, resulting in a reliable segmentation. From the simple segmentation, the coverage percentage of each of the biofouling categories was calculated based on the pixels in the image. The coverage percentages can stand along to provide information on the degree of biofouling, or it can be utilized to calculate a FRR according to the European Chemical Agency standard [4]. The results indicate that, utilizing image analysis to determine the coverage percentages for the FRR produces more objective and repeatable results, compared to using the coverage percentages obtained from the visual inspection. In addition, the evaluation procedure is substantially less time consuming. A standardized image analysis model would not require any interpretation of guidelines from the inspector, in terms of categorising the biofouling or determining the coverage percentages. Thereby, a better basis for comparison between different FCC could be achieved, which is important in the development of new eco-friendly FCC.

References:

1. D. M. Yebra, S. Kiil, and K. Dam-Johansen, “Antifouling technology - Past, present and future steps towards efficient and environmentally friendly antifouling coatings,” *Prog. Org. Coatings* 50, 75–104 (2004). ..
2. G. A. Hopkins and B. M. Forrest, „A preliminary assessment of biogouling and non-indigenous marine species associated with commercial slow-moving vessels arriving in New Zealand.“ *Biofouling* 26, 613-621 (2010)
3. M. L. Pedersen, C. E. Weinell, B. Ulusoy, K. Dam-Johansen, „Marine biofouling resistance rating using image analysis.“ Accepted in Journal of Coatings Technology and Research (2022)
4. European Chemicals Agency, “Transitional Guidance on the Biocidal Products Regulation - Transitional Guidance on Efficacy Assessment for Product Type 21 Antifouling Products,” Tech. Rep. May (2014).

Comparative investigations of UV curing coatings using LED- versus Hg-lamps

R. Nothhelfer-Richter, K. Weber, A. Krug

Fraunhofer Institute for Manufacturing Engineering and Automation IPA, Nobelstr. 12, 70569 Stuttgart/D

UV curable coating materials are interesting for various fields of application due to their rapid curing process and their special properties of the resulting polymer films.

In recent years, the performance of UV-LEDs regarding power and price has been steadily improved [1]. Thus, UV-LEDs became an attractive alternative to conventional mercury UV-lamps. Compared to Hg-lamps, the emission spectra of UV-LEDs are quasi monochromatic. Therefore, matching the radiation source and the recipe of the coating material is a crucial question for UV-LED applications [2].

In a current research project, relevant parameters of the UV-curing process have been investigated with special consideration of the UV-LED technology. The emission of UV radiators including their spectral energy distribution was measured. In this context, different measurement devices were compared as well. Devices without spectral resolution are able to control the stability of UV equipment, but their absolute values depend strongly on the type and manufacturer of the device.

The transmission and absorption properties of coating components like initiators, reactive thinners, and binders were measured and correlated to the emission spectra of the UV radiators. The recorded absorption spectra of the initiators are in good agreement with their technical data sheets. By exposure to UV-light, the reacting initiators change their absorption characteristics.

On the other side, data of binders and reactive thinners regarding UV absorption are poorly available. Most of these components showed a high UV absorption below 350 nm. Therefore, UV-B and UV-C radiation are less suitable for curing in the depth of the film.

Curing on the surface and in the depth of the film have been evaluated using model formulations. Matching of the emission spectra of the radiators and the absorption spectra of the initiators is necessary, but not sufficient, to achieve sufficient crosslinking.

Subsequent artificial weathering of the cured samples aims at investigations into possible changes like post-curing, embrittlement or yellowing.

References:

[1] Tache, F., Bourrousse, C. (2018) UV-Led has become dominant in many UV curing applications, *European coatings*, https://www.european-coatings.com/Raw-materials-technologies/Technologies/LED-has-become-dominant-in-many-UV-curing-applications

[2] Weber, K., Nothhelfer-Richter, R., Krug, A. (2019) UV-Härtungsleitfaden für organische Beschichtungen. 83. Lacktagung, Erfurt

Modern Weathering Testing of Paints and Coatings: Fluorescent UV & Xenon Arc

A. Francis, A. Giehl, B. Tobin, S. Fowler, D. Duecker, B. Reis

Q-Lab Corporation, Westlake, OH, (USA), agiehl@q-lab.com

Accelerated weathering testing using fluorescent ultraviolet (UV) and xenon arc test chambers is widely used to evaluate the effects of weathering on the physical and chemical properties of inorganic and organic coatings. These test architectures use different methods to accelerate and simulate the sunlight, heat, and water that coatings are exposed to outdoors. Fluorescent UV testers reproduce just the damaging UV portion of the sunlight spectrum while xenon arc testers reproduce full-spectrum sunlight. Both provide elevated specimen temperatures, though xenon arc instruments simulate higher temperatures for dark specimens because of the infrared (IR) light they deliver. Fluorescent UV test apparatus give a realistic, accelerated moisture attack via hot condensation, while xenon arc chambers deliver water via controlled relative humidity and water spray. In this presentation, these modern test instruments will be compared and contrasted.

The majority of accelerated weathering testing in fluorescent UV and xenon arc test chambers is conducted according to international standards from organizations like ISO and ASTM International, as well as specifications written by major original equipment manufacturers (OEM's). Many different standardized test cycles are available for testing but most are based on general cycles developed decades earlier that were subject to the technical limitations of test instruments at the time they were written. Accelerated weathering testing of coatings introduces special testing requirements that other industries may not face, including the need to evaluate resistance to cracking, blistering, and delamination. Specialized test cycles, especially those emphasizing the role of water in material degradation, can offer improved correlation with outdoor and in-service performance. This presentation will examine research results from some major test protocols used in today's laboratory weathering testing of coatings and highlight those most suitable for coatings.

Keywords: accelerated weathering, xenon arc, fluorescent UV, testing

Self-healing through Diels-Alder Reaction in uretdione based powder coating

N. Farshchi[1,3], M. Gedan-Smolka[2], B. Voit[2,3]

[1]Leibniz-Institut für Polymerforschung Dresden (IPF), Hohe Str. 6, 01069 Dresden, Germany, farshchi@ipfdd.de
[2]Leibniz-Institut für Polymerforschung Dresden (IPF)
[3]Technische Universität Dresden, Fakultät Chemie und Lebensmittelchemie, 01062 Dresden, Germany

The main reasons for powder coating development were environmental, ecological and energy related concerns. Powder coatings are potential alternatives for solvent-based coatings due to cost-efficient manufacturing process and application, little or no volatile organic content. In addition, economic benefits, being eco-friendly, elimination of hazardous waste and availability of raw materials along with accepted properties for various application are among their advantages. Taking into account the main concerns, self-healing abilities of powder coatings will be the next step toward elimination of the cost and pollution caused by mechanical impact of coatings. In addition, the recycling as an option would be available for thermoset powder coatings in this case.
In this study, we utilized a single molecule, which is prepared with Diels-Alder cycloaddition reaction and retro Diels-Alder cleavage reaction as a self-healing additive to achieve self-healing ability in the powder coating system. This powder coating system is based on uretdione cross linker and OH-polyester resin. Coatings prepared through melt mixing of components in lab mixer, milling, sieving and then application on the aluminium substrate through electrostatic spraying method. To illustrate the role of self-healing additive, various concentration (5 and 10 %wt.) in combination with different curing temperatures (120 and 160 °C) were applied.
Gradient oven were used to determine both suitable curing temperatures and also conventional coating tests such as wave scan, gloss, thickness measurements and impact test.
In order to determine the self-healing ability of the resulting films several characteristic methods such as Differential Scanning Calorimetry (DSC) coupled with optical microscope, Thermogravimetric Analysis (TGA), mechanical properties measurement with nanoindentation method, rheological measurements, Nuclear magnetic resonance (NMR), Micro surface test, Scanning electron microscopy (SEM), were applied on the samples.
Results revealed not only self-healing and recycling ability of samples but also significant drop in the curing temperature of the powder coating system compared to virgin system. Samples can undergo infinite numbers of self-healing process by applying elevated temperatures (120 °) about 13 min.

References:

[1] Feng, L., Yu, Z., Bian, Y., Lu, J., Shi, X. and Chai, C. Self-healing behavior of polyurethanes based on dual actions of thermo-reversible Diels-Alder reaction and thermal movement of molecular chains. Polymer. 124, 48-59, (2017)

[2] Prasad, L.K., McGinity, J.W. and Williams III, R.O. Electrostatic powder coating: Principles and pharmaceutical applications. International journal of pharmaceutics. 505 (1-2), 289-302, (2016)

Population balance equation model of deaggregation process carried out in high–shear dissolvers

R. Krzosa[1], Ł. Makowski[1], W. Orciuch[1], R. Adamek[2]

[1] Warsaw University of Technology, Poland

[2] ICHEMAD–Profarb, Gliwice, Poland

email address of corresponding author: Lukasz.Makowski.ichip@pw.edu.pl

Suspension particles deagglomeration takes place in dedicated devices, such as grinding mills and mixers equipped with high shear impeller. This process can be divide into three stages. First is powder wetting, that allows create a suspension, second is deagglomeration, breakage particles on smaller parts and the last step is adding substances that prevent agglomeration and creating lumps [1]. Particle size distribution influence on product properties therefore it is important to control the process to obtain satisfactory particle size. Second important parameter in particle breakage is energy consumption. Those parameters decides about final product quality and costs.

In this work deagglomeration process carried out in mixers were considered. Authors focused on influence of impeller geometry influence on breakage process, especially on final product size and energy consumption. To investigate this device work computational fluid dynamic (CFD) methods were applied. Due to high rotational speed of this devices k-ε turbulence model were used. To predict liquid surface deformation during process volume of fraction (VOF) multiphase model were applied. To solve population balance equations Direct Quadrature Method of Moments (DQMOM) were used. This method replace particle size distribution function by several number of Dirac's deltas, with some weights each, which have the same statistical moments as original function [2]. This method gives explicit equation for changes abscissas and weights of deltas in time. Therefore it allows investigate particle size distribution changes during the process. CFD methods allows to calculate important process parameters, such as mixing power, shear stresses, mixing time, fluid circulation, dead zone occurrence, particle size evolution etc. To verify simulations results experiments of deagglomeration process of titanium dioxide carried out in batch mixers equipped with several geometries of impellers were investigated.

It was carried out in laboratory scale which consist on 10 litres cylindrical tank equipped with impeller and control and measurement system. The obtained good agreement of theoretical and experimental results confirmed the validity of the applied mathematical models. The method makes it possible to determine the suitable process conditions for variable rheology nanosuspensions in order to obtain a product with the desired properties.

[1] Xie, L., Rielly, C. D., Ozcan–Taskin, G., Break–Up of Nanoparticles Agglomerates by Hydrodynamically Limited Processes. Journal of Dispersion Science and Technology 29, 573–579, (2008)

[2] Marchisio, D. L., Fox, R. O. Solution of population balance equations using the direct quadrature method of moments. Journal of Aerosol Science 36, 43–73, (2005)

Modelling of wetting & dispersing additives

A. Okkel*, I. van Kamp*, P. Jeschenko#, V. Kravchenko§

*: BYK-Chemie GmbH, Wesel, (Germany), e-mail: Andreas.Okkel@altana.com
#: DWI Aachen, Aachen, (Germany)
§: Moscow State University, Moscow, (Russia)

Wetting & dispersing additives are essential for an efficient stabilization of pigments and fillers in coatings. As there are lots of variables on the structural design of W&D additives, their development is usually a compromise between various possible chemical candidates and a reasonable development time. The tools of computational chemistry are not widely used in the chemical industry yet. In this study, we seek to evaluate the possibilities of chemical modeling for the development of W&D additives. The focus was to predict the ideal pigment affinic group for a specific pigment as well as the optimal microstructure for a set additive chemistry through modelling.[1] For computational modelling and molecular calculations a carbon black was selected, the pigment surface was characterized and a model for calculations established. W&D additives contain in general a solvent affinic and a pigment affinic part. To find the best fit for the pigment, a plurality of potential pigment affinic model compounds was evaluated. The interaction energies between particle surface and model compounds were calculated via molecular dynamics. Also model components of the solvent affinic parts of the final polymer were included in the affinity calculations. The solubilities of the model components were calculated via the Askadskii method in water and 1-Methoxy-2-propylacetate (PMA). To compare calculated and experimental data, model components were synthesized and their interaction energy with the carbon black surface was determined with different methods. Calculated and measured data were plotted against each other to validate the applicability of the computational method. For the best candidates, a course grain simulation of the polymers was performed to evaluate the affinity to the pigment surface. Finally, the affinity to the surface was investigated with different methods and tested in a coating formulation with synthesized model polymer analogues. Calculated and experimentally determined interaction values showed similar trends and identified several structures of being highly efficient in pigment adsorption. Additionally, the course grain simulations for the polymers indicated a trend for the pigment stabilization depending on the microstructure of the polymers. Even though the calculated values do not necessarily match the experimental values, the overall tendency seem to be aligned. The proof of concept seems successful and could provide a significant step forward in the digitalization of additive development. Nevertheless, there is still a long way to go to implement such methods into daily business.

Acknowledgements:
Prof. M. Möller, DWI Aachen; R. Vinokur, DWI Aachen; Dr. A. P. Vogt, formerly DWI Aachen; Prof. I. Potemkin, Moscow State University
We thank the ALTANA AG for the intellectual and financial support. Special thanks to Dr. Jürgen Omeis.

References:

[1] Perelstein, O. E., Ivanov, V.A., Möller M., Potemkin I. I. Designed AB Copolymers as Efficient Stabilizers of Colloidal Particles. *Macromolecules.* 43, 5442 – 5449, (2010)

Optimization of wet and dry hide in Architectural paints

Markus Vogel, Oliver Peters, Ellen Reuter

Evonik Resource Efficiency GmbH, Goldschmidtstr. 100, 45472 Essen, (Germany), e-mail: markus.vogel@evonik.com

For interior wall paints formulators are trying to ensure best possible hiding performance of the white pigment, which typically is titanium dioxide. Since this is a rather costly raw material, it is desired to minimize the usage level. This is typically limited by the impact that the reduction of the pigment amount on wet and dry hiding power of the formulation has. Moreover, the hiding properties are strongly influenced by the overall filler package in the formulation.

Specialty extenders enhance the yield of wet and dry hiding generated by the pigment by ensuring the optimal distribution/spacing of the titanium dioxide particles. Nevertheless, architectural paints are complex formulations: The addition of specialty fillers can result in changing the PVC of the formulation which in turn alters the morphology of the paint film. While low to medium PVC formulations typically have closed films, high PVC systems are porous. Binders can have very different monomer compositions and emulsifier systems, dispersing additives are used to improve the wetting & stabilization of particles and pigments as well as other ingredients, such as defoamers are needed.

To assess the interplay of all these ingredients and factors in this complex matrix can be very laborious. Different techniques, such as ladder studies, Design of Experiment setups or modelling approaches can help to narrow down the playing field.

But also lean testing regimes help to efficiently characterize the formulations. In this paper a new efficient optical way to test wet hide properties is shown via image analysis of wet coating films. Results from these classical laboratory set-ups can be augmented by automated testing results for generating recipes of efficiently optimized wall paint formulations.

ZEFFLE-GKD-210: A water-borne 2-component fluorinated polymer dispersion for highly chemically resistant and long-lasting coatings

Dr. M.Lemke, Dr. V. Kuznetsov, M.Farooq

Daikin Chemical Europe GmbH, martin.lemke@daikinchem.de

Providing low or VOCs-free components is one of the main global requirements for raw material suppliers in the coatings industry these days. Furthermore, the performance compromise in terms of long-term weatherability as well as durability of protection against corrosion, chemicals or dirt accumulation should be avoided. Such technical requirements are fulfilled by a high performing solvent borne 2K-coating using fluorinated FEVE-resin cross-linked with a suitable isocyanate (F-polyurethane). However, the presence of organic solvents restricts the use of these coatings especially for on-site applications. Therefore, introduction of FEVE-resin for waterborne-coating technology, which would offer comparable performance while allowing the reduction of VOC, remained a challenge until recently. [1][2]

DAIKIN is a one the leading suppliers of fluorochemicals in the world. Its solvent-borne fluoropolymer product line ZEFFLE GK covers the need for 2K-formulations in many industrial applications since 1990. As a producer of environmentally friendly products, DAIKIN sets very high sustainable development goals (SDGs) in all its business units. In order to meet them while fulfilling environmental regulations for heavy duty applications, DAIKIN has developed a new water-borne fluorinated polymer called ZEFFLE-GKD-210. Being a polyol, it can be cross-linked with polyisocyanate or melamine type hardeners at temperature from ambient up to 220°C. Chemically, it is copolymer of Tetrafluoroethylene (TFE, the monomer that is most popularly-used globally) and several Hydrocarbon olefins (HC), with pendent OH groups available for crosslinking. TFE part of the polymer delivers the characteristic fluororesin property of excellent durability against weathering, chemicals and moisture, while the HC part ensures compatibility with other coating components and results in film formation, gloss and adhesion to the primer/ substrate.[3]

Laboratory scale test results of a 2K white coating based on ZEFFLE-GKD-210 have proven its excellent chemical resistance against harsh chemicals like MEK, diluted acid or alkali solutions (e.g. 5 mass%). At the same time, the accelerated weathering (QUV-B) results of 1000 hours indicate a highly durable gloss and color retention properties as compared to traditional coatings. The determined performance values are equivalent to the reference solvent-borne 2K fluorinated coating within deviations brought by equivalent OH/NCO ratios. These results clearly demonstrate that ZEFFLE-GKD-210 can be used potentially in architectural, industrial, and automotive industries for both exterior and interior applications.

References:
[1] Best, K. Squiller, Dr. Edward. 2-Component Polyurethane Topcoats. PCI *https://www.pcimag.com/articles/88254-2-component-polyurethane-topcoats*
[2] Melchiors, M. Sonntag, M, Kobusch, C, Jügens, E. Recent Recent developments in aqueous two-component polyurethane (2K-PUR) coatings. *Progress in Organic Coatings*. 40 (1-4), 99-109, (2000)
[3] Pelz, S. Mantegazza, V. Coatings based on fluoropolymers: outstanding performance because of their chemistry. Pitture e Vernici - *European Coatings – Formulation*. 3, 10-15, (2018)

Laser-based processing of polymeric protection coatings for lightweight applications

Marius Dahmen (ILT)[1], *Dr. Hendrik Sändker (ILT)* [1], *Dr. Anna Buling (ELB)* [2], *Dr. Christian Vedder (ILT)* [1], *Jörg Zerrer (ELB)* [2]

[1] *Fraunhofer Institute for Laser Technology (ILT), Steinbachstraße 15, 52074 Aachen, Germany, e-mail: marius.dahmen@ilt.fraunhofer.de*
[2] *ELB Eloxalwerk Ludwigsburg, Neckartalstraße 33, 71642 Ludwigsburg-Neckarweihingen, Germany, e-mail: info@ceranod.de*

Using lightweight materials in automotive and aerospace applications is still an increasing trend, due to their high potential in terms of cost and energy efficiency. However, the alloys often do not meet the necessary requirements for corrosion, friction and wear protection. One solution to this problem is the modification of surfaces by protective coatings. Polymeric coatings based on high performance polymers such as polyether ether ketone (PEEK) show the potential to meet the necessary requirements due to their excellent properties in terms of temperature resistance as well as corrosion and wear protection.

Conventional oven-based processing of micro-particulate PEEK coatings has the major disadvantage of heating up the coating material as well as the entire substrate to the same temperature, which is in case of PEEK the melting temperature of about 340 °C. This results in a significant reduction in strength of temperature-sensitive components. An innovative approach to form a dense and adhesive protection layer on temperature-sensitive components is the melting of micro-particulate PEEK layers using laser radiation. Laser treatment overcomes the drawback of conventional oven-based processing, due to the rapid heating and cooling rates that allow the required temperatures in the film to be reached while reducing the thermal impact on the substrate material. This results in new challenges regarding the wetting behavior of the comparatively cold substrate surface with a hotter PEEK melt which can be met using an innovative laser-based process strategy.

The current investigation deals with the development of a laser process for the production of dense polymeric protective layers on temperature-sensitive components. This includes the determination of the optical properties of the coating materials and a selection of suitable laser beam sources. The investigation focuses on the influence of the substrate temperature on the adhesion properties as well as the active control of the substrate temperature while simultaneous melting the PEEK layer. In addition, various additives were tested to further increase the tribological properties of the coatings. As a result, it is possible to produce dense and adhesive polymeric protective coatings that exceed the performance of conventional coatings in terms of frictional properties.

Physical and Chemical Drying of Diverse Coatings Comprising Green Photocatalysts and Alternative Absorbers covering the UV, Blue and NIR Range with New LED Techniques

Bernd Strehmel[a)], Zhijun Chen[b)]

[a)]Niederrhein University of Applied Sciences, Institute for Coatings and Surface Chemistry, Department of Chemistry and Institute for Coatings and Surface Chemistry, Adlerstr. 1, D-47798 Krefeld, Germany. [b)]Northeast Forestry University, Key Laboratory of Bio-based Materials, Hexing Road 26, 150040, Harbin, China

Green photocatysts based on carbon dots[1] are introduced to show the initiation of photoinitiated polymerization in combination with LEDs emitting either blue, green or red. These materials can be derived from food waste and principally depict an alternative to petrochemical based photoinitiators keeping the price at a competitive level. Some work is in progress to make carbon dots absorbing in NIR to close the spectral coverage. Near-infrared radiation (NIR) uptakes in some industrial applications operating on Industry 4.0 standards an important function although this technique can be seen in its infancy considering the coating sector. Nevertheless, the significant lower scattering of photons compared to UV light can be seen as one main benefit[2,3].

Recently, development of photopolymer formulations based on radical and cationic crosslinking has started using a **H**igh **T**hroughput **F**ormulation **S**creening line resulting in development of artificial intelligence for coating applications. Functionalized cyanines covering their absorption between 700-1000 nm have become attractive absorbers/sensitizers in combination with new NIR-LED devices facilitating selective excitation at 808 nm, 850 nm, 930 nm, and 1000 nm. Several systems comprising an absorber and radical initiator were evaluated regarding their capability to crosslink multifunctional (meth)acrylic derivatives. Some NIR-sensitizer also successfully initiate cationic photopolymerization experiments using oxiranes in combination with functionalized cyanines with new onium salts. Moreover, these systems also possess potential to initiate physical processes such as drying of dispersions or melting of powder coatings. Examples demonstrate how the combination of physical (melting, evaporation of water) and chemical (photopolymerization) drying facilitate solid film formation.

Additionally, photochemical syntheses were pursued with NIR sensitizers based on a photo-ATRP protocol using Cu(II) catalysts at the ppm scale. This gave access to block copolymers with narrow dispersity comprising UV absorbing moieties[4].

References (few selected examples)**:**

[1] Kütahya, C.; Wang, P.; Li, S.; Liu, S.; Li, J.; Chen, Z.; Strehmel, B., Carbon Dots as Promising Green Photocatalyst for Free Radical and ATRP-Based Radical Photopolymerization with Blue LEDs. *Angew. Chem. Int. Ed.* **2020,** accepted. Open access.

[2] Strehmel, B.; Schmitz, C.; Cremanns, K.; Göttert, J., Photochemistry with Cyanines in the Near Infrared: A Step to Chemistry 4.0 Technologies. *Chem. Eur. J.* **2019,** *25* (56), 12855-12864. Open access.

[3] Schmitz, C.; Pang, Y.; Gülz, A.; Gläser, M.; Horst, J.; Jäger, M.; Strehmel, B., New High-Power LEDs Open Photochemistry for Near-Infrared-Sensitized Radical and Cationic Photopolymerization. *Angew. Chem., Int. Ed.* **2019,** *58* (13), 4400-4404.

[4] Kütahya, C.; Schmitz, C.; Strehmel, V.; Yagci, Y.; Strehmel, B., Near-Infrared Sensitized Photoinduced Atom-Transfer Radical Polymerization (ATRP) with a Copper(II) Catalyst Concentration in the ppm Range. *Angew. Chem., Int. Ed.* **2018,** *57* (26), 7898-7902.

Barrier properties of high performance PMMA-silica coatings on AA7075 and AA2024 alloy

P. Hammer, A. Trentin, M. C. Uvida

Sao Paulo State University-UNESP, Institute of Chemistry, Araraquara/SP, Brazil, peter.hammer@unesp.br

Organic-inorganic coatings are a promising alternative to replace primers containing a chromate conversion layer on aluminum alloys. The key feature of hybrid materials is their highly cross-linked nanostructure that acts as an effective diffusion barrier against corrosive species [1,2]. This work reports a detailed investigation on the barrier properties of PMMA-silica coatings prepared by combining the sol-gel process with the radical polymerization of methyl methacrylate (MMA) and cross-linking agent 3-(trimethoxysilyl)propyl methacrylate (MPTS) followed by the hydrolytic condensation of tetraethoxysilane (TEOS) and MPTS. Using dip-coating, few micrometer thick hybrid coatings were deposited on the AA7075 and AA2024 aluminum alloys. Five independent methods were used to determine in periodic intervals the water uptake during an immersion period of more than 1000 days in 3.5% NaCl solution. The uptake values were calculated from electrochemical impedance spectroscopy (EIS) data using the Brasher−Kingsbury relation [2] and from the high-frequency impedance values using the linear rule of mixtures [3]. The water uptake values were compared to those determined for free-standing films obtained for equal exposure intervals using gravimetry, thermogravimetry, and FTIR measurements. To study the structural effects of the water uptake, X-ray photoelectron (XPS) and ATR-FTIR spectroscopies were used to study the composition and bonding structure of the surface region of the coatings before and after immersion, while cross-sectional information was obtained using scanning electron microscopy (SEM) and energy-dispersive X-ray spectroscopy (EDS) maps. EIS assays showed excellent anticorrosive performance of the coatings with an essentially unchanged low-frequency impedance modulus of more than 1 GΩ cm^2 after exposure of 1000 days. SEM/EDS images confirmed an unaffected homogeneous cross-sectional morphology and XPS detected a reduction of the surface concentration of silicon due to hydrolysis induced leaching of the silica phase. The final water uptake results showed for all methods very low values of less than 3%. This remarkable barrier properties can be explained by the highly reticulated hybrid structure resulting in a very slow progression of the uptake zone of only a few nm/day towards the coating/alloy interface.

References:

[1] Trentin A., Gasparini A. de L., Faria F.A., Harb S. V., dos Santos F.C., Pulcinelli S.H., Santilli C.V., Hammer P. Barrier properties of high performance PMMA-silica anticorrosion coatings, Prog. Org. Coatings. 138, 105398 (2020).

[2] Dos Santos F.C., Harb S. V., Menu M.J., Turq V., Pulcinelli S.H., Santilli C.V., Hammer P. On the structure of high performance anticorrosive PMMA-siloxane-silica hybrid coatings, *RSC Adv. 5* 106754–106763, (2015).

[3] Brasher, D.M.; Kingsbury, A.H. Electrical Measurements in the Study of Immersed Paint Coatings on Metal. I. Comparison Between Capacitance and Gravimetric Methods of Estimating Water-Uptake. *J. Appl. Chem.* 4, 62–72, (1954)

[4] Kremer F., Schönhals A. Broadband dielectric spectroscopy, *Springer Science & Business Media* (2002).

Outstanding Durability in Superhydrophobic Coating Achieved by Sol-Gel Process

Rana Rafiei Hashjin, Zahra Ranjbar, Hossein Yari

Institute for Color Science and Technology, Faculty of Surface Coating and Novel Technologies, Tehran, Iran, ranjbar@icrc.ac.ir

Properties of hydrophobic and superhydrophobic surfaces such as self-cleaning, anti-bacterial, anti-fouling and anti-icing has made the list of their applications impressive which includes textile and coating industry, construction, medicine, energy and photovoltaic devices, automobile and aerospace industry. In this research, a robust cost-effective route is introduced to achieve an exceptionally durable superhydrophobic coating with environmental-friendly chloro/fluoro-free precursors. Tetraethyl orthosilicate, Methyltriethoxysilane and Triethoxyoctylsilane precursors and fumed silica nanoparticles in water/ethanol mixture were processed through sol-gel method in presence of acidic and basic catalyst, separately. Final coating was prepared by spraying sol on glass substrate and thermal treatment in 100℃. Static contact angle and sliding angle of water droplet on coatings were measured by a homemade apparatus, to monitor wettability of samples. Surface topology was investigated by scanning electron microscopy and atomic force microscopy to study surface roughness and developed structures. Fourier-transform infrared spectroscopy and X-ray spectroscopy were also employed for chemical analysis. Contact angle of 160° with near zero sliding angle was achieved by adjusting sol-gel parameters and coating application route. Best results were obtained by spraying alternate layers of acidic and basic sol. Observations and obtained results in this study are in good agreement with fundamental discussions regarding reaction mechanisms during sol-gel process. As the superhydrophobicity uses have already reflected in our life, durability became a matter of issue which addresses preserving hydrophobicity behavior during their practical utilization. To this end, durability tests were carried out. Remaining superhydrophobic after being in accelerated weathering condition (8 hrs UVA irradiation at 60℃ followed by 4 hrs condensation at 50℃) for 1500 hrs was considered as an outstanding long-term durability. Peeling tape test and slurry erosion were also conducted as complimentary experiments in which coating remained hydrophobic after 17 runs and 71 minutes of being under test, respectively. Such an excellent durability in addition to scalable inexpensive fabrication route render this coating as a promising candidate for outdoor applications.

References

[1] D. Levy, M. Zayat, eds., The sol-gel handbook, 2015.

[2] S.P. Dalawai, M.A. Saad Aly, S.S. Latthe, R. Xing, R.S. Sutar, S. Nagappan, C.S. Ha, K. Kumar Sadasivuni, S. Liu, Recent Advances in durability of superhydrophobic self-cleaning technology: A critical review, Prog. Org. Coatings. 138 (2020) 105381.

36 Years of paint and resin innovation: from invention to innovation

Dr. Jaap Akkerman, Retired chemist and NVVT board member, Goes, The Netherlands

New "inventions" can be totally new chemistry, major or even small improvements of existing technology. They may very well not be new: they often exist many years as "sleeping inventions" in the literature, in own laboratory or in patents that were not successful in our daily business. An invention is not an innovation: the innovation is a successful invention. This is a process – "innovation circle" – that can last long and fail easily.

The reason may very well be that inventions came too early or were too complicated to reproduce. They could also be taken over by alternative, better, less complex technologies. Or worse: be stopped, by the reluctance of the management or even the ignorance of the R&D people. Another cause is often the underestimation of the complexity: too many raw materials, process variables, parameters but also our customers or changing legislation. The failing invention can be discovered again in a different context like changing markets/management or advancing time or simply by studying existing reports or literature.

In this presentation the process of innovation will be discussed as I experienced this in the last 36 years. From an "invention", via proving and development of the concept, validation, upscaling and fine-tuning and ending up with testing and proving the proper application at the customer, leading to acceptance and sales in the market. It is based on experiences of several short and long lasting projects where I was involved, at Sikkens, Synthese, Akzo, AkzoNobel, AkzoNobel Resins, Nuplex and Allnex. The stodgy practice often led to hick-ups and even complete failures. In some cases a re-birth of the invention to innovation happens, even after decades.

Using both examples outside our business as well as inside our business, patterns can be recognized that led to this presentation and that I would like to share with you now that I am pensioned. No rocket science: hands-on experience and tips, for what it is worth. Useful for the future? Who knows.

Coatings binder systems from high functionality bio-based seed-oil derived resins

Dean C. Webster, Raul Setien, Samantha Silbert, Arvin Yu

North Dakota State University, Fargo, North Dakota, USA dean.webster@ndsu.edu:

As one approach to addressing issues of sustainability, the exploration of bio-based materials in coatings continues to accelerate. An overall challenge is to meet or exceed the performance properties of current coating systems as well as to have economic viability and demonstrated sustainability. This can be accomplished through careful consideration of polymer design and molecular architecture as well as considerations of the sources and availability of raw materials.

In a series of studies, we have been exploring derivatives of epoxidized sucrose esters of fatty acids as high functionality bio-based resins that can yield high performance materials when crosslinked. A focus has been on epoxidized sucrose soyate (ESS). By esterifying unsaturated fatty acids to sucrose, an eight-membered polyol, a compact molecule can be formed that has a high number of double bonds. These double bonds can be readily converted to epoxy groups, yielding a molecule having a large number of functional groups along with a moderate viscosity. Using ESS directly, it has been found that coatings that are 100% bio-based can be made by curing ESS with bio-based di- and tri-functional acids and that cured coatings having a good combination of hardness and flexibility can also be prepared using short chain acid oligomers as crosslinkers [1].

Derivatives of ESS can also be prepared by further functionalization reactions of the epoxy group. Several different functional groups can be employed to use as crosslinking sites. Carbonated sucrose soyate (CSS) is made by reacting ESS with carbon dioxide [2]. CSS can be cured with multifunctional amines to yield polyurethanes without the use of isocyanates. Another non-isocyanate polyurethane system is generated by making a carbamate functional derivative of ESS and curing it with multifunctional aldehydes. The coatings are hard, solvent resistant, and have good adhesion to substrates. Acrylate functional ESS can be cured with amines via the aza-Michael reaction to yield coatings that have high hardness, good solvent resistance, and good adhesion [3]. A new method of epoxidation using benign components is also being explored in order to make the resin more “green.”

References:

1. Ma, S., D.C. Webster, and F. Jabeen, *Hard and Flexible, Degradable Thermosets from Renewable Bioresources with the Assistance of Water and Ethanol.* Macromolecules, 2016. **49**(10): p. 3780-3788.
2. Yu, A.Z., et al., *Catalyzed non-isocyanate polyurethane (NIPU) coatings from bio-based poly(cyclic carbonates).* J. Coatings Techn. & Res., 2019. **16**: p. 41-47.
3. Paramarta, A. and D.C. Webster, *The exploration of Michael-addition reaction chemistry to create high performance, ambient cure thermoset coatings based on soybean oil.* Progress in Organic Coatings, 2017. **108**: p. 59-67.

Reducing CO2 emissions with 100% Bio-Based and fully segregated raw materials – Additives and building blocks based on renewable raw materials for sustainable paints and coatings

Abstract

Reducing greenhouse gas emissions is undoubtedly one of the greatest challenges for society and for industry in this century.

With the European Green Deal of December 11th, 2019, the European Union has made climate neutrality, the circular economy and an environment free of hazardous substances the main goal and calls for the more sustainable use of scarce fossil resources. The chemical industry also wants to make its contribution by using renewable raw materials. One of the important raw materials for emulsifiers, wetting and dispersing agents and building blocks for resins are ethylene oxide derivatives.

There are first developments to address those needs. Clariant has introduced its new VITA range consisting of raw materials based on renewable raw materials and originating from a segregated supply chain. Those products are almost identical to fossil-based raw materials; yet they provide a carbon emission saving in comparison to the conventional counterpart enabling the reduction of Scope 3 emissions in the supply chain.

It is expected that the European Union will expand their carbon taxation scheme soon by including scope 3 emissions in the range of carbon taxation.

Therefore, bio-based and renewable products such as the VITA range will help companies to reduce their carbon footprint. In view of further upcoming emission taxation, these products will contribute to reduce costs in comparison to conventional fossil-based products.

Address

Vincent Fritzemeier
Business Development Manager Coatings EMEA
BU Industrial & Consumer Specialties

Clariant International Ltd.
Rothausstr. 61
4132 Muttenz 1
Switzerland

Mail: vincent.fritzemeier@clariant.com

Eco-friendly In-Can Protection of Paints in a Sustainable Environment

Timo Schock, David Tierney

Sanitized AG, Burgdorf, Switzerland, David.tierney@sanitized.com

In a world where everything is possible and available, the way we behave, consume, and produce is changing drastically. We`re becoming greener, more eco-friendly, and looking for sustainable solutions. These effects have also reached the paint industry for years and require new approaches in the selection of raw materials and production. After general functionality, sustainability (including eco-friendliness) has become the second priority of a product design to be accepted in the market. But what about the environmental challenges of paints? Have they changed? Not really. At least not in the world of microbes. They`re omnipresent in a paint's environment at all times: In natural based raw materials, during manufacturing, and in the application environment.

The waves of sustainability and environmental protection have hit the paint and biocide industries hard, creating a conflict of interest. It is a question of which angle we want to look at the issue from and what we want to focus on. Is it sustainable if a significant amount of paint spoils and is disposed of due to a lack of biocides? Probably not. Is it eco-friendly to use biocides in formulations? Well, it depends. Is it even possible for a biocide to be sustainable and environmentally sound? Compromises on both issues are rarely accepted.

Yes, we believe it`s a matter of responsible use and decision making. Products such as Sanitized BZ, based on 5% BIT and 5% zinc pyrithione can work at active dosages of less than 50ppm each. This requires good manufacturing practices, a clean plant and proper handling after production and during application. The same is true for other combination products such as BIT + bronopol or sometimes just BIT. However, the most important change is that biocides are individual formulation solutions rather than generic assurances, that can be applied in all water-based formulations without further testing.

In an efficacy study, conducted in accordance to ASTM D 2574, we found that even 0.05% (25 ppm BIT and 25 ppm ZPT) provided good product protection, if the product was not frequently exposed to high contamination. [1] The study showed that it`s much more important to protect the paint from severe contamination spikes and give it enough time to recover from the contamination than to add high levels of biocide. The study also challenges the industry's common practice of testing performance with a so called 7-day cycle in-can challenge test. Technically: It is more important to control microbes consistently over the long term than to kill them very quickly. Conversely, the results of this study also show that spoilage can only be prevented by high biocide loads if the contamination is too severe or occurs too frequently within a short period of time. Since microbes are invisible, it is important to measure and control them. Once their growth gets out of control, it is very difficult to assess the risk of product failure, even at high biocidal dosage.

We`d be glad to guide you through this process with our experience and find the best available solution for your products in the in-can and dry-film field.

[1] Sanitized AG, Internal Study, in-can challenge testing with Sanitized BZ and Nexiva CT115 under different conditions (2021)

Vacuum Inline Dispersion saves Titanium Dioxide and inducts dust free

Dr Hans-Joachim Jacob

ystral gmbh maschinenbau + processtechnik

Titanium Dioxide causes up to 45% of the raw material costs in architectural paints and up to 20% in average industrial and decorative coatings. 95% of al RAL colours contain TiO2.

Due to several huge price increases during the past years the reduction of Titanium Dioxide became one of the main targets for cost savings in paint industry. The use of substitutes or enhancers is only one way. A much more effective dispersion of the already applied Titanium Dioxide is the better way.

Traditional dissolver discs do not reach a fully colloidal dispersion. A new in line dispersion method creates 1000 times higher shear rates and significantly higher concentrated energy density in combination with vacuum expansion of the powder. This way between 2 and 12% Titanium Dioxide are saved with same or even better quality of the final product.

Same hiding power and colour strength are reached with less Titanium Dioxide. The undertone is not influenced since particles are not smaller – but colloidal dispersed and better distributed.

To achieve these savings not only the machine but also the optimum process sequence is important. The recipe may remain the same, but the order of addition is new.

Another topic is safe handling: Powdered titanium dioxide is supposed to be classified as carcinogen category 2 by inhalation (H351) in future. The vacuum induction provides a safe and dust free handling, a closed induction directly into the liquid.

LIFE-BioPaint and LIFE-WB BioPaint: demonstration of innovative sustainable processes for resin and paint production

Marcello Vitale, Guido Sangiovanni, Ilaria Lombardi

IVM Chemicals Srl, Bareggio, (ITALY), m.vitale@ivmchemicals.com

Sustainability is no longer a «nice to have» but a necessity that every industry must now pursue. It has also become increasingly clear that it must be assessed multi-dimensionally using a whole-system approach. To minimize the environmental impact of industrial products, product innovation must meet process innovation at all levels.

Sustainable improvements over state of the art for a paint or coating are, for example:

1. **product composition**: high bio-based and low VOC content, low health and pollution risk
2. **production process of coating and its components**: low energy draw, low emissions, low waste, high efficiency, low variability so that there are no rejects
3. **product application**: as above, with low variability application on a suitable substrate, such as wood from sustainably managed forests
4. **end-of-life management**: long useful life and the enabling of well-managed collection and disposal processes.

IVM Chemicals, a global leader in wood coatings, led LIFE-BioPaint and is now leading LIFE-WB BioPaint, EU-funded projects aimed at piloting and demonstrating strategies to implement this vision. In particular, the former project addressed UV-cured resins and coatings, the latter targeted WB systems.

The European Union's LIFE Program fosters the development of actions, strategies and technologies that improve environment, biodiversity and climate. Its funding of these projects is making it possible to build on the one hand the most advanced paint production line in the world, on the other hand to make available in water the best coating resins until now available only in organic solvents.

LIFE- BioPaint's paint production pilot line

- is automated, closed circuit and without any VOC emission
- generates minimal waste thanks to optimized cleaning and uses much less energy than current standards
- delivers components with maximum precision via double metering
- affords optimized dispersion by rheological control and can produce identical lots of a few kgs or several tons
- adds no variability to the products and ensures consistent top quality.

The last point is especially important for biobased paints, whose raw materials often show more variability than their petrochemical versions; indeed, new biobased resins were also developed for production on this line of UV-cured, 100% solid coatings that ensure the lowest environmental impact at application as well as the longest service life on the most sustainable of substrates, wood.

In **LIFE-WB BioPaint** we are building a pilot line for dispersing polymers of low hydrophilicity in water via extrusion, a technology applied for the first time to alkyd and polyester polyols. Such resins of high biorenewable content are produced by IVM for solvent coatings that are seen as the best for wood applications upon isocyanate crosslinking. We expect to be able to produce WB paints of comparable quality by using the same resins in water, driving the SB-to-WB substitution process for a great decrease in VOC and GHG emissions.

The presentation will detail both the overall line designs and the individual innovative solutions that are being integrated and implemented.

Barrier properties of an epoxy novolac resin system under conditions of high pressure and high temperature: Influence of inorganic fillers and its morphology

Narayanan Rajagopalan, Claus Erik Weinell, Søren Kiil.
CoaST Research Centre, Department of Chemical and Biochemical Engineering, Technical University of Denmark, DTU, Building 229, 2800 Kgs. Lyngby, Denmark.
naraj@kt.dtu.dk

The present study explores the barrier properties of amine-cured epoxy novolac (EN) resin systems under high pressure high temperature (HPHT) conditions, defined by the US code of Federal Regulations [1,2], as a pressure rating greater than 100 bars and/or a temperature greater than 170 °C. The aim has been to investigate the effects of different inorganic fillers (i.e. concentration, and morphology) on the performance behaviour of EN systems before and after HPHT exposure.Due to high consumptions and fast depleting resources, the increase in demand for energy is pushing the oil and gas industry to explore and produce in geological formations that show abnormally high pressures and high temperatures (HPHT). Corrosion under HPHT is predominant and all applications in the HPHT zones are under severe attack from the extremely corrosive environment. Conditions of high pressures and temperatures in the presence of hydrocarbons, seawater and carbon dioxide gas can increase corrosion rates with unexpected degradation pathways [2]. Owing to the extremes, process equipment, wells, tanks, and pipelines in the HPHT zones are often lined with highly cross-linked epoxy coatings for substrate durability and corrosion protection. The morphology and structure of fillers can impact the barrier properties of such coating systems and here we studied the effects of aluminium-silicates with flake structure, hollow-spheres with spherical structure and meta-silicates with acicular structure. HPHT conditions of the oil and gas industry were simulated encompassing three phases in a chamber namely; a gas phase that includes a mixture of nitrogen (N_2) and carbon dioxide (CO_2), para-xylene (PX) as the hydrocarbon phase and 3.5% NaCl solution as the seawater phase at high pressures and high temperatures. EN systems incorporated with different fillers were analysed before and after HPHT exposure. Characterization techniques viz. Scanning electron microscopy (SEM), Energy Dispersive X-Ray (EDX), Fourier Transform Infrared Radiation (FTIR) and Differential Scanning Calorimetry (DSC) were used to understand the structure relation properties of the fillers with respect to its corrosion resistance. Using SEM, the changes in surface topography and cross-section images were mapped and showed the dispersion state of fillers and its condition after HPHT exposure. EDX showed the elemental composition at all areas and cross sectional line mapping analysed the depth of penetration of HPHT components in to the coating system. FTIR analysed the chemical changes in the resin crosslinking before and after HPHT exposures. Lastly, DSC allowed understanding the change in glass transition temperature, supporting the changes in corrosion resistance and the overall coating system. The results demonstrate the characteristic barrier properties of each of the fillers with respect to degradation pathways under HPHT conditions. It was concluded that each HPHT phase interacts differently with the crosslinked network and owing to the filler morphology, the coating system provides different torturous paths and thermal stability minimizing the interaction of corrosive HPHT components.

References:

1. Shadravan, Arash and Mahmood Amani. "HPHT 101: What every engineer or geoscientist should know about high pressure high temperature wells. *SPE Kuwait International Petroleum Conference and Exhibition*. Society of Petroleum Engineers, 2012.
2. DeBruijn, G., Skeates, C., Greenaway, R., Harrison, D., Parris, M., James, S. & Wutherich, K. High-pressure, high-temperature technologies. *Oilfield Review*, *20*(3), 46-60, (2008).

Synthesis of new organosilicon derivatives for coating applications

Adrian Franczyk[1], Jakub Szyling[1,2], Kinga Stefanowska[1,2], Tomasz Sokolnicki[1,2], Jędrzej Walkowiak[1]

[1]Center for Advanced Technology, [2]Faculty of Chemistry, Adam Mickiewicz University in Poznan, Poznan, (Poland)
adrian.franczyk@amu.edu.pl

Functional silanes and polyhedral oligosilsesquioxanes (POSS) are one of the most intensively studied reagents dedicated to the synthesis of a wide spectrum of novel hybrid materials including those for coating applications. The importance of silanes in this field is well established while the importance of POSS, due to their excellent physicochemical properties, increases more and more, and is reflected in significant numbers of scientific papers and patents. The rigid inorganic siloxane core endows POSS with good mechanical properties, thermal stability, UV resistance, and low surface energy, while the soft organic substitutes afford good reactivity. These advantages make them perfect surface modifiers and polymer components which affect the surface energy and roughness, and can improve targeted physicochemical properties of the resulted coatings.[1]

However, new and effective preparation methods of novel silanes and POSS (and materials based on them) are still highly required. Therefore, in this communication, the development of synthetic methods of silanes and POSS bearing a wide spectrum of reactive groups will be presented. These approaches include but are not limited to the industrially applied catalytic hydrosilylation of C=C and C≡C bonds, allowing obtaining of molecules with high yield. Resulted compounds are characterized by unique structures and physicochemical properties and can play different roles in the preparation of novel, advanced coating materials.[2] The presentation will be also devoted to the application of selected POSS in the synthesis of hybrid poly(methacrylate)s (*via* atom transfer polymerization (ATRP) and free radical polymerization (RP)) which can be considered as valuable precursors of protective coatings.[3]

Acknowledgments: The authors acknowledge the financial support from the National Centre for Research and Development in Poland: LIDER/6/0017/L-9/17/NCBR/2018.

References:

[1] a) Chen, F., et al., Polyhedral Oligomeric Silsesquioxane Hybrid Polymers: Well-Defined Architectural Design and Potential Functional Applications. Macromol. Rapid Commun., 40 (17), 1900101, (2019).

[2] For example see: a) Stefanowska, K., Franczyk, A., *et al.,* An effective hydrosilylation of alkynes in supercritical CO_2 – A green approach to alkenyl silanes. Journal of Catalysis. 356, 206-213 (2017). (b) Walczak, M., Stefanowska, K., Franczyk, A., *et al.,* Hydrosilylation of alkenes and alkynes with silsesquioxane $(HSiMe_2O)(i\text{-}Bu)_7Si_8O_{12}$ catalyzed by Pt supported on a styrene-divinylbenzene copolymer. Journal of Catalysis. 367, 1-6 (2018). (c) Walczak, M., Franczyk, A., *et. al.*, Synthesis of Bifunctional Silsesquioxanes $(RSiMe_2O)_{\sim 4}(R'SiMe_2O)_{\sim 4}Si_8O_{12}$ *via* Hydrosilylation of Alkenes. Organometallics. 38, 3018−3024 (2019).

[3] Alexandris, S., Franczyk, A., *et. al.*, Polymethacrylates with Polyhedral Oligomeric Silsesquioxane (POSS) Moieties: Influence of Spacer Length on Packing, Thermodynamics, and Dynamics. Macromolecules. 48 (10), 3376–3385 (2015).

New waterborne binders with IKEA R2 performance in clear, satin and pigmented coatings

Jurgen Scheerder, Sam Huizing, Remy Dollekens, Michel Luijsterburg, and Remko de Zwart

Covestro Coating Resins B.V., Sluisweg 12, 5145 PE, Waalwijk.

jurgen.scheerder@covestro.com

There is a continuous trend in the coatings industry towards coating systems with reduced coalescent demand yet increased coating performance. This brings significant challenges with respect to the design of waterborne binders. To tackle these challenges, multiphase polymer dispersions have been developed that combine hard (high Tg) and soft (low Tg) polymer phases. By varying the ratio between the two polymer phases parameters like hardness and coalescent demand can be tuned. This technology is by now well established. However, it suffers from a limitation that there is a direct link between the coalescent demand and the amount and Tg of the hard polymer phase. Increasing the amount of hard polymer phase typically improves coating properties like chemical resistances and mechanical properties, but at expense of an increasing coalescent demand. To break through this dilemma a new design of multiphase polymers is required. In this paper we will introduce novel morphology controlled binders that show an excellent balance between the chemical resistances (IKEA R2 in clear and pigmented formulations) and the mechanical performance at <80 g/l VOC. In addition, these binders are easy to formulate. Two key aspects will be highlighted. The first aspect relates to the coalescent demand. Coalescent efficiency studies have been done to understand the effect of differences in particle morphology on the coalescent efficiency. These studies show that by selecting the proper particle morphology coalescents are used very effectively resulting in a significant reduction in coalescent demand. The final VOC demand of the clear satin and pigmented satin formulations are 67 g/l and 84 g/l respectively, which is well below the IKEA R2 threshold of 130 g/l. The second aspect deals with the influence of the high Tg monomer building blocks. The effect of the monomer structure on the polymer backbone and its effect on the film formation temperature is discussed and molecular modelling in employed to gain understanding on the effect of the monomer structure on the polymer backbone configuration. Next the performance of the novel morphology controlled binders is discussed. Although many of the current waterborne binders in the market show IKEA R2 chemical resistances in clear glossy, upon matting and/or pigmenting they show a significant drop in chemical resistances to well below IKEA R2 level. The novel morphology controlled binders do not show this drop in chemical resistances upon matting and/or pigmentation. Hence the novel morphology controlled binder deliver IKEA R2 performance in clear and pigmented coatings from satin to high gloss. As such these binders, which do not require crosslinking, further enables the shift from solventborne to waterborne binders.

Will the Smart Paint Factory Eliminate the Need for Lab Personal Within the Next 5 Years?

U. Stalmach, R.J. Wörheide, H. Hustert

ORONTEC GmbH & Co. KG, (Germany), e-mail: ulf.stalmach@orontec.com

Digital tools, e.g. the Industrial Internet of Things IIoT, Big Data, and most notably Artificial Intelligence AI are buzz words connected by most people with two things: a huge boost in productivity – and the fear of loosing their jobs! The rather conservative Coatings Industry is no stranger to this, quite the opposite.

While it is true that the way we will work in the future will change (for some more, for some less drastically), this presentation aims at describing how these tools can help free valuable, highly educated experts from doing repetitive "dummy" work – change that surely every coatings specialist actually yearned for for years!

We will elaborate on several examples of how working digitally can free up valuable specialist time, thus implementing our vision of The Smart Paint Factory. The first example covers corrosion test panel evaluation. While traditionally the test panels have to be scratched to free them from loose coating and rust before they can be analyzed, we describe an automized test method that can 1) measure up to 25 samples automatically overnight, and 2) leave the sample surface unmodified from mechanical force due to using a thermo-optical method for measuring.

Next, by using a liquid color measurement device, we are able to drastically save sample preparation time (and energy and material, for that matter) while gaining increased accuracy and reproducibility of the measured data compared to measuring on dry samples. Even more, this fast and accurate method enables the quality control of incoming pigments and pigment pastes, as it does not rely on film forming samples. The variation of pigment quality can be factored into the formulation, thus allowing the right-first-time production of batches much closer to the desired color than with the most often used ("standard") procedure of a static formulation, followed by correction to hit the target color.

When thinking about the transfer of information, this usually is done according to organizational or IT system rules, resulting in breaks of the information flow and thus intransparency. If the flow of information can be arranged along the process instead, relevant data is available at all time at the actual spot of the process. The key word here is process automation. Who would not love to skip procedures like "read out a hand measuring device, type the values into an Excel sheet, then send it somewhere by email" – and subsequently fumble around for hours to retrieve the data when someone asks about it? Instead, we will present an example how to retrieve all relevant data online within the flow of the process, independent of the underlying IT architecture of the client.

Considering these examples to answer the question in the title of this presentation, the answer is of course a resounding No: lab personal will always be necessary for the foreseeable future. However, all available digital tools will certainly change the way a lab technician will perform his/her routine work in the years to come. Certainly, the examples described above will result in more time available for doing brain work fitting much more to the specialist education and experience they obviously have. Computers are excellent in doing repetitive routine work – let them do what they can do best, so lab technicians can do what they can do best.

High Throughput support for optimization of high quality pigment preparations

Ellen Reuter, Claudia Bramlage, Dr. Nina Hoppe

Evonik Resource Efficiency GmbH, Goldschmidtstr. 100, 45472 Essen, (Germany), e-mail: ellen.reuter@evonik.com

Optimization of additive packages for high quality pigment preparations can be a very complex task: although the systems typically consist of just a few raw materials, the choice of grades or dosages can be critical to achieving the desired performance or avoiding negative side effects. Typically, in order to get to an optimized formulation a huge number of lab trials is needed making the development of pigment concentrate lines for demanding applications a complex project. In addition many coating producers are looking for optimized raw material baskets and want to narrow down the number of additives used across different pigments in a colorant line.

In order to address challenges in the fields of coatings and inks with speed, efficiency and precision, Evonik has developed a fully automated lab scale high throughput system for production of coating formulations, their application and characterization together with the Swiss manufacturer Chemspeed. Because of the afore mentioned complexity the system is especially suited to assist in the development of optimized high quality pigment preparations.

Examples of results from different projects on high quality pigment preparations are shown – as well as comparisons to laboratory test results since the grinding of pigments to the desired coloristic properties has to be considered.

Typically in projects on high quality pigment preparations, the HTE system

- produces a broad variety of concentrates with different additives & combinations as well as different dosages

- characterizes rheology of the liquid formulations & homogeneity after storage,

- tints the pigment preparations into a variety of different base systems

- coats substrates via bar coater or spray application & allows curing at room temperature or at elevated temperature,

- characterizes the finished coatings regarding e.g. film thickness and coloristic properties as well as gloss & haze,

- takes images of the dried films & carries out image analysis

With its broad applicability the new, fully automated high-throughput system enables Evonik to cut down significantly on time spent searching for the best possible additive package for a formulation or chemical structure of new products - in doing so, Evonik accelerates not only its own pace of innovation, but also that of its customers.

Coating degradation and rust creep assessment - A comparison between a destructive method according to ISO 12944 and selected non-destructive methods

Huichao Bi[a,*], *Claus Erik Weinell*[a], *Benjamín Santos Varela*[b], *Sergio González*[b], *Álvaro Rodríguez Ruiz*[b], *Kim Dam-Johansen*[a]

a. *CoaST, Department of Chemical and Biochemical Engineering, Technical University of Denmark, 2800 Lyngby, Denmark*

b. *Fundación Centro Tecnológico (CTC) C\ Isabel Torres nº1, Parque Científico y Tecnológico de Cantabria, 39011 Santander, Spain*

**Corresponding author. E-mail: hubi@kt.dtu.dk.*

Organic coatings have been widely applied for the protection of marine structures from corrosion. Usually, the coating systems for the marine environment are specified to last more than 20 years. However, degradation of the coating system starts already as it is commissioned and repair and maintenance are usually needed before the planned service lifetime of the structure. Rust creep is one of the most important criteria for the prequalification of an anti-corrosive coating system for offshore applications, as established in ISO 12944-9. This evaluation method is destructive in nature which means that one panel only gives one reading of rust creep because the coating beyond the rust area must be removed. Non-destructive methods allow for the evaluation of coating performance in an more efficient and cost-effective way where the same coated panel can be tested over time so that the early changes/breakdowns of the coatings and the propagation of the rust creep can be detected and monitored, which will lead to a better understanding of the mechanism behind. In this present work, three methods based on ISO 12944-9, optical 3D Profilometry and Scanning Acoustic Microscopy (SAM) for the rust creep assessment are compared and results show that the optical 3D Profilometry and SAM can act as the non-destructive test methods providing more efficient and accurate rust creep evaluation.

Novel waterborne acrylic technology for sustainable basecoat formulations

Wouter M.J. Kloosterman*, Robert Harrer‡

*Allnex Netherlands BV, Bergen op Zoom, NL, wouter.kloosterman@allnex.com
‡Allnex Austria GmbH, Graz, AT, robert.harrer@allnex.com

Driven by green innovation targets like volatile organic compound reduction, waterborne technologies for automotive coating systems were already invented back in the late 80s, and waterborne basecoat formulations for car refinish and automotive OEM systems found their first applications in the early 90s. Since then, the scope of sustainability development goals for coating systems has further broadened from volatile organic compound reduction to complete assessment of all the environmental aspects associated with a product over its life cycle. At the same time, plastic has become one of the key materials required for the structure, performance, and safety of automotive vehicles in recent years, with growth in plastic consumption being driven by light weighting trends for fuel efficiency and consequently lower greenhouse gas emissions.[1]

The established waterborne basecoat formulations require innovation and continuous improvement to anticipate on above mentioned trends. Allnex developed a new and unique alkali swellable acrylic resin with hydroxy-functionality for water-borne basecoat formulations that meet the modern basecoat requirements like low temperature cure and excellent adhesion to a variety of plastic substrates without the need for a primer or pre-treatment. The resin presented here is developed such that it is free of substances of high concern, does not contain any solvents other than water and due to its strong pseudo-plasticity after neutralization, is an ideal binder for metallic basecoat formulations. Combined with a polyurethane dispersion, the new acrylic development can be used for two-component metallic coating formulations fulfilling the demanding requirements for automotive OEM interior coatings such as hot water and chemical resistance. Due to the lower thickener demand, those formulations also show an improved storage stability with respect to viscosity and pH-value. Due to its high hydroxy value, application of the novel binder in one-pack refinish basecoat formulations can result in a better distinctness of image and improved inter-coat adhesion between for example the basecoat(s) and the applied clearcoat as well. This presentation will provide insights in the synthetic route used to produce the new acrylic binder and will demonstrate the above mentioned advantages in two-component metallic coatings.

References:

[1] P. Khemka, Plastics in the Automotive Industry – Which Materials Will Be the Winners and Losers? www.nexant.com, Feb 27 (2019)

More Durability of Coatings by Enhancing Scratch and Abrasion Resistance

Marco Heuer, Roger Reinartz, Fabian Eichenberger

Evonik Resource Efficiency GmbH, Essen, Germany,
e-mail: marco.heuer@evonik.com

Resistance to scratches and abrasion is one of the mayor challenges and one of the most important properties for modern coating systems [1].
When talking about property improvement, formulators have different approaches to achieve a higher level of scratch and abrasion resistance. Based on the experience background of different formulators the solutions can be achieved by totally various options for sudden coatings systems.
But before coming up with a recipe solution regarding an extra addition of raw materials it is necessary to make a deep dive into an important question upfront: What is the necessary testing method? Some of the common methods are so differently, that an anticipation of results is hard to transfer.
In this study four major dimensions are stressed for a reliable and objective answer: five coating systems, different raw materials to enhanced the properties, variation in loading level and five different testing methods.
The used formulations are two one-component-wb-coatings, one two-component-PUR-wb-coating, one 2-component-PUR-sb-coating, one sb-high solid UV-curable coating. So these choices covered a lot of the common state-of-the-art coating systems for industrial use.
In different loading levels inorganic materials like precipitated silica, structure modified silica, functionalized nanosilica of the portfolio are added in test formulation. Next to these products also polyester and ketone co-binders with relatively high glass transition temperature are tested as well as functionalized and non-functionalized polyether-polysiloxane surface active additives [2].
The testing methods had been a summary of tests used in automotive industry, film and foil industry or consumer electronics. Next to Taber Abrader also Taber Shear test, Crockmeter, Martindale and Martens-hardness are used to create results. Some of these tests are pure penetration into the coating matrix, some have an additional movement, so more a tendency to scuffing or indenting stress [3].
As general results based on these four dimensions, a ranking of different raw materials in various levels for the used formulations was made.

[1] Research and Markets: Abrasion Resistant Coatings Market by Type (Metal/Ceramic, Polymer), End-Use Industry (Oil & Gas, Marine, Power Generation, Transportation, Mining, Construction), Region (North America, APAC, Europe, MEA, South America) - Global Forecast to 2024 (2019)
[2] www.coating-additives.com
[3] M. Heuer: Testing of Scratch- and Abrasion Resistance of Coating Systems for Permanent Protection of Plastics-83. Lacktagung der Fachgruppe Lackchemie, Erfurt (11.09.2019 - 13.09.2019)

Achieving the next level of durability without reformulation – New hardener for ambient curing high heat resistant coatings

M. Hallack, M. Heuer, S. Herrwerth

Evonik Operations GmbH, Essen, Germany

e-mail: markus.hallack@evonik.com

Heat resistant coatings are used to form a permanent multiresistant layer on metallic substrates after application and curing for permanent protection and aesthetic aspects. Next to fire places also many industrial applications require a persistency against temperatures above +180°C. A formulation of a coating resistant to high temperature is state-of-the-art based on silicone resins (Methyl-Phenyl-Silicone Resins). Usually these resins can be cured in two ways: via oven curing (mostly 30 min 250°C) or via a newer way of hydrolysation-condensation-curing by using alkoxy functional groups of the resin. The oven curing is done by putting the coated substrate after flash-off time for 30 min at 250°C in the oven, taking them out for 1 h at room temperature for cooling down before it can be used for handling or packaging. The hydrolysation-condensation-curing means that it will be applied on the substrate and cured at ambient temperature usually over night before handling or packaging can start. Lower temperatures or lower humidity will slow down the curing reaction and longer the curing time. A strong customer need is the combination of the positive topics of curing at room temperature and short timeframe. With the completely new developed hardener based on a special Silane/Siloxane technology it is the first time to achieve that goal. Next to that no reformulation work on the coating composition is necessary with the usage of this new reactive cobinder. The resistance to heat exposure remains on the same high level as the stateof-the-art curing process. It can be used for all kind of Methyl-Phenyl-Resins as well as for Silicone-Hybrid resins. A ladder study has made upfront to evaluate the best curing time-performance ratio. In lab studies this non-stochiometric behaviour ranges from 5-25% addition level. By using this patented technology it is possible to achieve direct-to-metal heat resistance coatings layers of >75μm fulfilling the requirements of corrosion protection class C5. Also the new co-binder offers benefits regarding 12 months storage, 100% active matter and a very low viscosity of around 10 mPas respective 10 cps which offers also the option to use it as a reactive dilutant without any VOC emission caused by solvents. So it helps to create high-solid-formulations as well as 100% solid content coatings. Get ready for the future: Achieving the next level of durability without reformulation.

Wetting of Polymeric Binders Derived from Traditional and Ionic Polymethacrylates

V. Strehmel, K. Knobloch, B. Strehmel

Institute for Coatings and Surface Chemistry, Niederrhein University of Applied Sciences, Faculty of Chemistry, Krefeld, Germany,
e-mail: veronika.strehmel@hsnr.de

Polymeric binders based on polymethacrylates have been commonly applied for various organic coatings.[1] The structure of these polymers can be varied from relatively non-polar materials up to polymethacrylates comprising highly polar ionic structural motifs. The latter were obtained by polymerization of ionic liquid monomers using traditional free radical polymerization or group transfer polymerization.[2]

Glass transition temperature of the polymers is important for the application temperature range of organic coatings. It strongly relates to the chemical structure of the polymers. The length of the main chain and the substituent at the ester group importantly affect the glass transition temperature of the traditional polymethacrylates. In addition to these structural elements, the size of the alkyl substituent at the cationic structure at each segment of the polymer chain and the mobile counter ions strongly affect the glass transition temperature of ionic polymethacrylates as well. Furthermore, the chemical structure of the polymethacrylates is not only important for their solubility in various solvents, it also influences the wetting behavior of organic coatings. Moreover, different drying methods of the polymer films were compared considering drying at ambient lab conditions on the one hand, and application of vacuum drying conditions on the other hand.

Contact angle measurements were carried out using various solvents (water, ethylene glycol, and toluene) that distinguish with respect to their surface tension.[3] Surface energy, is similar for both types of polymethacrylates obtained either by free radical polymerization or group transfer polymerization. However, both the alkyl substituent at the cationic moiety and the mobile anion influence the surface tension. The latter composes of a disperse part and a polar part that are differently influenced by the polymethacrylate structure.

References:

[1] Wicks, Z. W., Jones, F. N., Pappas, S. P., Wicks, D. A., Organic coatings science and technology, 3rd Ed., John Wiley & Sons, Inc. (2007).

[2] Strehmel, V., Berdzinski, S., Ehrentraut, L., Faßbender, C., Horst, J., Leeb, E., Liepert, J., Ruby, M. P., Senkowski, V., Straßburg, P., Wenda, A., Strehmel, C., Application of ionic liquids in synthesis of polymeric binders for coatings, Prog. Org. Coat. 89, 297-313, (2015).

[3] Wu, S., Polymer interface and adhesion, M. Dekker, New York, (1982).

Recent achievements in silicone-based flame resistant coating materials

M. Zielecka, Ł. Pastuszka, L. Jurecki

Scientific and Research Centre for Fire Protection National Research Institute (CNBOP-PIB), Józefów, (Poland), e-mail: mzielecka@cnbop.pl

Modern coating materials must possess excellent properties to meet the requirements of the most innovative applications used in electronics, vehicles, household appliances, measuring and medical equipment and others. A very important parameter that should be taken into account is the flame resistance of the applied coating materials because it affects the safety of innovative technological solutions. Silicones and hybrid polymeric materials containing silicones are one of the most economical solutions in the field of coatings due to the effect of the unique combination of very good thermal insulation and resistance with surface properties.
The results presented in most relevant works, including patents, concerning flame resistant silicone-containing coating materials will be presented. Thermal degradation of silicones occurring according to two different mechanisms depending on the combustion conditions will be presented. In contrast to the running smoothly thermal depolymerization process resulting in cyclosiloxanes formation, the thermo-oxidative degradation process proceeds with the release of CO, H_2O, CH_2O, CO_2, methanol and residual amounts of formic acid as degradation products of organic groups attached to the polysiloxane chain, and in the final stage SiO_2. Based on the study of thermo-oxidative degradation processes, it can be stated that the stability of polysiloxanes depends primarily on the nature of the organic groups constituting substituents in the siloxane chain. The stability of these substituents can be ranked as follows: C_6H_5 > ClC_6H_4 > $Cl_3C_6H_2$ > $Cl_2C_6H_3$ > $CH_2 = CH$ > CH_3 > C_2H_5. The second factor influencing the thermal stability of polysiloxanes is their molecular weight.

The flame resistance of polysiloxanes systems based on their thermal degradation data including the effect of additives as well as the flame resistance of hybrid silicone-containing coating materials will be discussed.

The effect of layered-silicate fillers on the flammability and thermal stability of silicone composites will be presented in detail. Based on the results of research carried out by different research groups it can be stated that the mechanism of flame retardance by these fillers is associated with the formation of a sintered layer of a silica-carbon composite in the course of combustion of the polysiloxane. Sintered layer formation is a complex process associated with a number of possible mechanisms enabling the growth of flame resistance thanks to the following effects: mass and heat barrier, labyrinth, steric and catalytic.

The most important innovative applications of these materials and future trends in application of silicones will be also discussed.

Anticorrosive coating materials based on Zn–particles

J. Lach, M. Kędzierski, E. Zakrzewska, A. Łukomska, J. Przybylski, W. Tokarz, J. Trzaskowska, K. Sylwestrzak, W. Sarna, K. Wróbel
Łukasiewicz Research Network – Industrial Chemistry Institute, Warsaw, (Poland), e-mail: jakub.lach@ichp.pl

Keywords: anticorrosive coatings, zinc-rich paints, electrochemical production of zinc nano/micropowder

One of the methods of protecting metal surfaces from corrosion are zinc-based paints. Their anticorrosive properties are the result of three main mechanisms: cathodic protection, barrier protection and adhesion [1]. This type of paints can be based on many different binders, e.g. epoxide, acrylic, polyurethane or silicate. The main goal of the conducted research was developing an optimal composition of paints based on different Zn particles and testing the performance of the prepared paints.

The influence of the method of production and characteristics of Zn particles on the properties of the paints was investigated by conducting a series of experiments. The performance of the paints was tested by measuring their hardness, adhesion to metal, rheological properties etc. Different methods of application were also employed, e.g. painting using multiple layers, application of an outer coating layer. A range of chemically curing and thermoplastic binders, both organic and inorganic was used and paints with satisfactory performance were obtained.

Additionally, a series of electrochemical experiments was carried out, which tested the resistance to corrosion of the paints. For the best developed compositions an increase in the duration of cathodic protection in comparison to selected commercial paints with similar binders was observed.

The completed research allowed to produce zinc-rich paint compositions based on zinc powders with good quality of the coating and application properties, as well as strong cathodic protection.

References:

[1] Zhang, X., G. Zinc-Rich Coatings, in: *Corrosion and Electrochemistry of Zinc*, Plenum, New York, (1996)

BIOFIRE Project– Advanced Coated Steels for Biomass Firing

K. Radwański,[1] R. Kuziak,[1] W. Cerri,[2] A. Chierichetti,[2] B. Zabala,[3] E. Fuentes,[3] R. Bayón,[3] S. Heckmann,[4] J. Meitz,[4] B. Corlu,[5] K. Gillner,[6] M. Aminyavari,[7] A. Landi,[7] R. Valle,[8] S. Lionetti,[8] A. Meduri,[8] R. Pileggi,[8] D. Donato.[8]

[1]Sieć Badawcza Łukasiewicz Instytut Metalurgii Zelaza Im Stanislawa Staszica, Gliwice (Poland). [2]Flame Spray Hungary Femipari Szolgaltato Es Kereskedelmi Kft, Szada (Hungary). [3]Fundacion Tekniker, Guipuzcoa (Spain). [4]Rwe Power AG, Essen (Germany). [5]Onderzoekscentrum Voor Aanwending Van Staal NV, Zelzate (Belgium). [6]Mannesmann Precision Tubes GmbH, Hamm (Germany). [7]Bono Energia S.p.A., Milan (Italy). [8] Rina Consulting - Centro Sviluppo Materiali S.p.A., Rome (Italy).

BIOFIRE project aims at increasing the service life of boiler tubes working in the harsh environment of opportunity fuels, such as biomass. Thanks to its ubiquitous nature, biomass (co)firing is currently being considered as a valid pathway to mitigate the adverse effects of fossil fuels (e.g., coal). As a consequence, the feasibility to employ or convert existing coal fired boilers for biomass (co)firing is a hot topic [1-2]. It is not an easy task: biomass is not only very different from coal but is also a heterogeneous, difficult, and complex fuel. Its (co)firing involves slagging, fouling and corrosion, low ash fusion temperatures, erosion, and abrasion [3-5]. To limit such phenomena, a promising solution – and the one adopted in the present project – is the application of protective diffusion coatings onto existing boiler steel (e.g., T12, T22). This solution ensures an application in the short term since the base material is certified with well-known performances. Moreover, diffusion coatings are available for both inner (steam side oxidation) and outer (fireside corrosion) tubes.

Aluminised diffusion coatings were designed, developed, and deposited on T12 and T22 base materials via pack, gel and slurry techniques. Coated samples were characterised via microstructural analysis (OM, SEM, XRD, GDOES, hardness), thermo-physical and mechanical tests (elasticity, corrosion, erosion, tensile, creep, ...), and in terms of workability (bending, welding). Results demonstrate the enhancement of the coated over the base materials.

References:

[1] Sahu, S.G.; et al. Coal-biomass co-combustion: an overview. *Renewable and Sustainable Energy Reviews* 39, 575-586 (2014).
[2] Sugiyono, A.; et al. Potential of biomass and coal co-firing power plants in Indonesia: a PESTEL analysis. *IOP Conf. Ser.: Earth Environ. Sci.* 963, 012007 (2022).
[3] Altobelli Antunes, R.; Lopes de Olivera; M.C. Corrosion in biomass combustion: A materials selection analysis and its interaction with corrosion mechanisms and mitigation strategies. *Corrosion Science* 76, 6-26 (2013).
[4] Vassilev, S.V.; et al. An overview of the composition and application Of biomass ash: Part 2. Potential utilization, technological and ecological advantages and challenges. *Fuel* 105, 19-39 (2013).
[5] Shao, Y.; et al. Ash Deposition in Biomass Combustion or Co-Firing for Power/Heat Generation. *Energies* 5, 5171-5189 (2012).

Silane functional polyisocyanates and prepolymers and their coating applications - A novel approach

I. Latorre, L. Woods, F. E. Golling, H.-J. Laas, Leverkusen/D

Dr. Irene C. Latorre, Covestro Deutschland AG, Kaiser-Wilhelm-Allee, Leverkusen/D

Polyurethane made of polyisocyanates and polyols are very well established for a wide range of coating applications due to their outstanding properties, such as excellent hydrolytic stability, UV-resistance, low temperature flexibility, and resistance to solvents and chemicals. The polycondensation of silane groups in the presence of moisture is another well-known crosslinking reaction for these applications. Polyisocyanates and polyurethanes additionally containing silane moieties combine the properties of both worlds in one molecule or prepolymer. Such isocyanatosilane building blocks and isocyanatosilane-containing prepolymers allow for the formulation of high quality polyurethane coatings with improved features, such as abrasion resistance. However such building blocks suffer from rather high viscosities, thus requiring extra solvents for thinning and leading to unfavorable volatile organic component (VOC) emissions. We will present a new concept for the synthesis of low viscous silane functional polyisocyanates and prepolymers and discuss their potential for different applications.

The effect of castor oil on properties of carboxy methyl cellulose based edible coating

Hossein Yahyaei, Nima Fakhraei, Parisa Oloub

Department of Polymer Engineering and Color Technology, Amirkabir University of Technology, P.O. Box 15875-4413, Tehran, Iran

Coatings industry could find a broad application in variety of industry. By sure, the application of coatings does not restrict to metal to prevent corrosion or making beauty of the house walls. One of the interesting area that coatings can play special role is application of edible coatings to procrastinate the fruit deterioration. Every year a large portion of foods, especially fruits and vegetables, from the time they are harvested until they reach to consumers, go through different kinds of deterioration.
Various factors are involved in food deterioration such as microorganisms, mechanical damages, time and light. Therefore, if we fail to protect the food, it can be damaged within hours or days. One way to prevent or delay the food deterioration is the use of edible coatings. The purpose of this study was to evaluate the effect of adding various concentrations of castor oil (2% and 4% v/v) on the properties of edible films based on carboxymethyl cellulose (CMC). Moisture content, water solubility, tensile strength, elongation at break, elastic modulus, water vapor permeability, optical and thermal properties and antioxidant activity of the films were examined. The results demonstrated that the presence of castor oil led to a decrease in moisture content, water solubility and mechanical strength of the films. The film containing 4% of oil showed the highest water vapor permeability. The optical properties measurement represented that all of the samples were transparent. By the addition of castor oil, the antioxidant activity of the films improved largely. Due to the presence of unsaturated double bonds in the castor oil structure, this oil had the ability of radical absorption and therefore antioxidant activity.
Thermal properties of the samples had also been investigated and it was determined that the effect of castor oil on the melting point was negligible. Finally, the performance of the coatings for protecting fresh apples was studied with measuring the pH value of apples and performing some field test.

References:

[1] *Han, J. H., Edible Films and Coatings: A Review. Innovations in Food Packaging. 24 (3), 239 – 262, (2005).*

[2] *Biswal, D. R., Singh, R. P., Characterisation of carboxymethyl cellulose and polyacrylamide graft copolymer. 57 (4), 379 – 387, (2004.)*

The effect of cobalt-free promoters on curing of unsaturated polyester resins at ambient temperature

Eliška Matušková [a], Jan Honzíček [a], Jaromír Vinklárek [b]

[a] *Institute of Chemistry and Technology of Macromolecular Materials, Faculty of Chemical Technology, University of Pardubice, Pardubice, (Czech Republic), email:st42098@student.upce.cz*

[b] *Department of General and Inorganic Chemistry, Faculty of Chemical Technology, University of Pardubice*

Our study is focused on behaviour of cobalt-free transition metal compounds as curing promoters for unsaturated polyester resins (UPR). The curing process is based on a free radical polymerization including transformation from the liquid state into the rigid structure. If the curing process is performed at ambient temperature, accelerator and initiator need to be added. As initiators, organic peroxides are commonly used, for example methyl ethyl ketone peroxide (MEKP). Accelerators, particularly complexes of transition metals, are effective in decomposing of peroxides, which serve as sources of free radicals [1]. With increasing environmental pressure on the producers, it is important to find a cobalt-free alternative for the sustainable future of polyester resins [2].

The catalytic power of potential promoters has been verified by determination of gelation time on a commercial polyester/styrene/methyl ethyl ketone peroxides system. In case of exothermic behaviour, we focused on a different situation during the curing of UPR. At thermally isolated system, the curing process was followed by a K-type thermocouple, which enabled to observe inhibition of the curing process at high concentration of promoters and verify their thermal stability. In case of isothermal conditions, consumption of styrene and unsaturated polyester double bonds was followed by real-time near-infrared spectroscopy. Subsequently, a kinetic model was established for precise quantification of catalytic effect for each promoter. Temperature of glass transition was determined by using thermomechanical analysis in order to verify the ability of tested promotors to reach similar degree of cure as a commercially used promotor based on cobalt carboxylates.

All tested promoters were proved to be capable of catalysing the curing process of the unsaturated polyester resin with a potential for future use.

References:

[1] Mleziva, J., Šnupárek, J. *Polymery: výroba, struktura, vlastnosti a použití.* 2, (2000)

[2] Leyssens, L., Vinck, B. Cobalt toxicity in humans–A review of the potential sources and systemic health effects. *Toxicology*. 378, 43–56, (2017)

This research was funded by Ministry of Education of the Czech Republic (UPA SGS_2022_005)

New Tools to Tune the Wettability of Multi-layer Coatings!

Dr. Guillaume W. Jaunky, Mark Heekeren, Dr. Marc Eberhardt

BYK-Chemie GmbH, Wesel, (Germany), Guillaume.Jaunky@altana.com

Modern coatings face a number of technical and commercial challenges. Coating additives have been used successfully as "problem solvers" in coating formulations and are becoming increasingly important to enable coatings formulators to reach ever-increasing performance requirements and to advance the technology.

Single-layer coatings have a range of applications in the paint and coating industries. However, there are an increasing number of applications where the demands for the finish are high and the properties provided by a single layer system are not sufficient. One way to surmount these issues is to use multi-layer coatings combining the attractive functions of several layers, each chosen to bring certain properties in the application. For example, an automotive finish is not only attractive, but also provides an important protective function. A multi-layer coating is a multi-component system and its surface/interfaces have a strong influence on its quality. Surface defects often occur during and after coating application, and degrade both the optical properties of the coating and its ability to protect the substrate. The most common defects are poor substrate wetting, non-optimal flow (orange peel) and crater formation. One very significant parameter affecting all these defects is the surface tension of the materials involved. More specifically, surface tension differences are the actual cause of these surface defects. The source of surface tension differences can originate within the system itself (solvent evaporation or the crosslinking reaction of the resin) or from external sources (overspray, dust particles, or contamination of the substrate).

Wetting is a crucial coating attribute, especially in waterborne systems as their surface tension is higher. Consequently, using the appropriate wetting agent becomes of upmost importance. Additives that migrate toward the coating/air or coating/substrate interface and tailor surface or interfacial tension can improve wetting. In this presentation, we will briefly discuss the conventional route to decrease the surface tension of the coating to be applied, and afterwards focus on a second route to increase the surface energy of an underlying coating by hydrophilic additives, which positively affects the wetting of the subsequent coating layer.[1],[2] Various unresolved issues with recoatability were the drivers and motivation for developing this second and unique opportunity to build up multi-layer systems. The additives we recently developed are multi-functional and combine good levelling, anti-crater and slip properties to the hydrophilic properties, a unique combination of effects!

References:

[1] Bessel, Michael., Jaunky, Guillaume., Heekeren, Mark. Primed for improvement: hydrophilic agent enhances levelling and intercoat adhesion. *European Coatings Journal*. (10), 27 – 31, (2014)

[2] Sulzbach, Horst., Heekeren, Mark., Eberhardt, Marc., Jaunky, Guillaume. Next-Generation Additives. *Coatings World*. (10), 50 – 54, (2018)

Functional coatings for sand printed moulds for casting of electrical insulation in energy distribution systems

A. Rybak, *A. Siwek*

ABB Corporate Technology Center, Kraków, (Poland), e-mail: andrzej.rybak@pl.abb.com

Energy consumption is continuously growing in the developing world, therefore an electrical engineering industry puts a lot of attention to the effectiveness of electrical devices. Thermosetting polymers, i.e. epoxy resins and liquid silicone rubbers, are commonly used as the main electrical insulation materials in the energy distribution systems (e.g. instrument transformers, sensors, bushings, embedded poles, gas insulated switchgears). Their common usage in the electrical applications is due to the superior electrical insulation properties and a very good mechanical and chemical resistance offered by these materials. As the manufacturing process a reactive moulding technology is often used, and two methods are mainly applied, namely: vacuum casting and automated pressure gelation (APG). In the both cases the metallic moulds are used in order to cast and cure the final products.

Design of the appropriate shape of the moulds used in casting of epoxy materials is essential in respect to proper initiation and propagation of curing fronts that influence the final quality of the cast products. Moreover, very often it is necessary to manufacture the moulds for prototyped designs and for urgent short-series production. In such cases it can appear that the traditional mould machining is too expensive, and in addition a long awaiting time cannot be accepted by the customers. Therefore, it is of importance to overcome such technical limitations and evaluate a novel approach based on application of an additive manufacturing technology for moulds fabrication.

The presented work will be focused on an unique 3D printing-based concept for manufacturing of the moulds used in manufacturing of electrical products with epoxy electrical insulation. The proposed approach involves a sand binder jetting technology allowing a very fast manufacturing process of complex moulds with high dimensional stability, and what is important for multiple casting purposes. This technology is mainly used in metal casting processes, but here the sand mould is used only once, since after metal solidification it is destroyed. Therefore several methods of mould mechanical reinforcement by means of impregnation were proposed [1]. Due to relatively high surface roughness of the printed sand, a novel concepts for dedicated functional coatings have been evaluated and tested in order to achieve a robust process and good material integrity. The applied coatings have improved both, the surface quality and the anti-stick functionality, and as a result the demoulding was facilitated. The results of the first successful industrial epoxy casting trials with the use of developed coatings will be also presented, and a plan for further coatings improvement will be discussed.

References:

[1] Bednarowski, D., Javora, R., Kmita, G., Matysiak, Ł., Rybak, A., Sekula, R. Method of preparation of sand casting moulds with a protective coating. *Patent WO2019231341A1*, (2019)

Designing hybrid nanocomposites based on molybdenum disulphide and carbon nanomaterials for enhanced photo-electrocatalytic properties

Zuzanna Bojarska, Marta Mazurkiewicz-Pawlicka, Łukasz Makowski

Faculty of Chemical and Process Engineering, Warsaw University of Technology, Warsaw (Poland), zuzanna.bojarska.dokt@pw.edu.pl

Designing and manufacturing new nanostructures with improved properties using modern techniques is a tempting prospect for nanotechnology. Molybdenum disulphide (MoS_2) is a widely used 2D nanomaterial. MoS_2 found various applications such as a dry lubricant, in hydrogen evolution reaction (HER) catalysis, hydrogen storage, and others. MoS_2 can be an excellent candidate for being combined with carbon nanomaterials (CNMs) to obtain new hybrid nanocomposites with outstanding properties including higher photo-electrocatalytic activity and improved rheological and tribological properties.

The aim of the conducted research was to develop of the facile and scalable method to produce hybrid nanocomposites based on MoS_2 nanoparticles deposited on carbon nanomaterials, such as graphene oxide (GO) and reduced graphene oxide (rGO). Synthesis of MoS_2/CNMs hybrid nanostructures was carried out in the semi-batch reactor and in the turbulent micromixer with coaxial geometry by wet chemical synthesis. Ammonium molybdate tetrahydrate $(NH_4)_6Mo_7O_{24}\cdot 4H_2O$ (HMA), citric acid $C_6H_8O_7$ (CA), and ammonium sulphide $(NH_4)_2S$ (AS) were used as the substrates of the reaction. Graphene oxide was prepared from graphite powder using modified Hummer's method. Reduced graphene oxide was obtained by the chemical reduction method using a 50% hydrazine solution. The nanocomposites were synthesized in different ratios of MoS_2:CNMs i.e. 50:1, 30:1, 10:1. Physicochemical analysis of obtained materials was carried out, using various analytical techniques: thermogravimetric analysis (TGA), X-ray diffraction (XRD), particle size analysis by laser diffraction (PSD), and scanning electron microscopy (SEM). Furthermore, the photo-electrocatalytic properties of the obtained materials have been checked.

The wet chemical synthesis in the micromixers is a proper and facile approach to produce MoS_2/CNMs hybrid nanocomposites. This method allowed to easily obtain a product with desired properties i.e. particles in the nanometric scale (~100 nm). The best results were obtained for the MoS_2:GO sample in the ratio 50:1, the average particle size was ~70 nm. Scanning electron microscopy images showed that the graphene flakes were thoroughly covered with MoS_2 particles. The addition of carbon materials to the structure also promotes the charge transport of MoS_2, due to their high electrical conductivity and large surface area, which is valuable for catalytic applications.

Acknowledgments:

This work was supported by the National Science Centre [No.2017/27/B/ST8/01382]

References:

[1] Cravanzola, S., Cesano, F., Magnacca, G., Zecchina, A., Scarano, D. Designing rGO/MoS_2 hybrid nanostructures for photocatalytic applications. *RSC Adv.* 6, 59001 – 59008, (2016)

[2] He, Z., Que, W. Molybdenum disulfide nanomaterials: Structures, properties, synthesis and recent progress on hydrogen evolution reaction. *Applied Materials Today.* 3, 23–56, (2016)

Deep matte effect – Novel precipitated silicas achieve enhanced performance

T. Klotzbach, M. Heuer

Evonik Operations GmbH, Essen, Germany,
e-mail: thomas.klotzbach@evonik.com

Precipitated Silicas are known for more than 40 years as excellent matting agents for paints and coatings. Softtouch coatings for plastics and wood are gaining more and more market share. Typical applications like consumer electronics, kitchen cabinets and automotive interior parts are just some examples for usage of softtouch coatings in our daily life.
The majority of industrial used softtouch coatings are served by solventborne and waterborne resin systems.
The rising design aspects of softtouch coatings require a smooth, fine structured surface as well as highest transparency.
With novel precipitated silicas of very low particle size it is possible to reach an excellent haptic effect combined with a high transparency to achieve freedom for formulators to create new colortone effects with special effect pigments or over galvanized plastic substrates.
To show the dependency of matting efficiency and the used binder system also results will be presented of waterborne and solvent coating systems.
Current challenges within waterborne formulations will be identified and possible solutions explained in detail.
Next to these topics also the test results of the influence of novel precipitated silica particles to other physical respective chemical properties will be shown for a complete picture to formulate proper softtouch coatings.

Hybrid alkyds, the glowing route to reach cutting-edge properties

M. Denis[a,b*], D. Le Borgne[b], S. Caillol[a] , C. Negrell[a]

[a] ICGM, Univ Montpellier, CNRS, ENSCM, Montpellier, France
[b] Lixol, groupe Berkem, 20 Rue Jean Duvert, 33290 Blanquefort, France
*Corresponding author: maxinne.denis@enscm.fr

Alkyd resins were polymerized for the first time in the mid-1920s and they quickly found a prominent place among coatings and paint binders in industrial field [1]. Indeed, they provide numerous advantages such as good thermal resistance, excellent gloss, fast drying time [2]. In the 1950's, the emergence of synthetic polymers with more specific properties for advanced applications weakened the position of the alkyd resins for the coating industry [3]. However, since the past decade, due to growing up environmental concerns a new researchs on alkyd resins were carried out. Indeed, alkyd resins are mostly composed of biobased raw materials which makes them more "environmentally friendly" than other polymers based on fossil ressources. in order to enhance alkyd properties, bridge the gap with new materials, and remain competitive in terms of applicative properties, other polymers such as epoxy, acrylates, polyurethanes or siloxanes have been used to form alkyd hybrids. Recently, our team highlighted the various stategies to enhance alkyd properties and develop alkyd hybrid polymers in a review article [4].

Recent researches have been focused on the modification of alkyd resins with an oligomer containing silicon for coating applications on wood sample. Silicon-modified polymers are well know to exhibit excellent thermal properties and provide outdoor properties. The results of the study have showed that silicon modified alkyd resins exhibited better coating properties such as faster drying time, higher gloss and hardness and allowed to determine the optimal content of oligomer. Moreover, other heteroatomes can be associated to silicon modified alkyd in order to provide a synergistic effect and bring new properties such as fire resistance.

Currently, due to low chemical and corrosion resistance, a novel hybrid alkyd-polyhydroxyurethane (PHU) has been investigated for metal coating applications. The coatings properties have been studied throught salt spray test and accelerated weathering tester, and exhibited better performance with the introduction of urethane functions. Moreover, the hybridization of alkyd resins with PHU have enhanced the adhesion properties of the coatings.

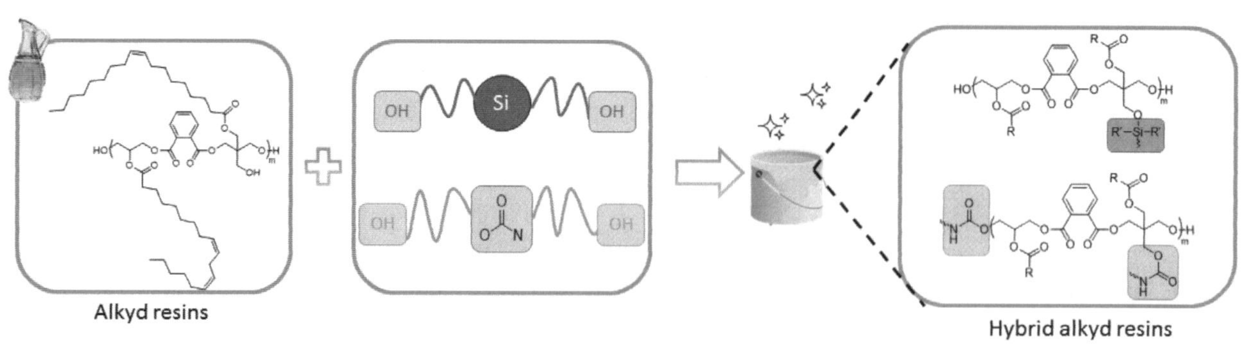

References

[1] Elliott, W. T., Alkyd resins, *Surface coatings*, **1993**, 76-109.

[2] Hofland, A., Alkyd resins: From down and out to Alive and Kicking, *Prog. Org. Coat.*, **2012**, 73 (4), 274-282.

[3] Manning, L., Moreillon, O., Hamacek, K., Opportunities for the alkydresin market, *Paint. Coat. Ind.*, **2019**.

[4] Chardon, F., Denis, M., Negrell, C., Caillol, S., Hybrid alkyds, the glowing route to reach cutting-edge properties ?, *Prog. Org. Coat.*, **2021**, 151, 106025.

Faster – Better – Cheaper: Recipe and Process Optimization

F. Kother, S. Jonas

TMC GmbH, Duisburg (Germany), office@tech-mc.de

Who doesn't know such situations: Many time-intensive small-volume products that reduce production capacity? A product portfolio that has become too large over the years and thus difficult to organize? Stockpiling of raw materials that are rarely used? Necessary rework to correct production fluctuations? Excessive forklift traffic in production and warehouses, or declining margins in traditional processes?

Often, product portfolios have grown over many years. They were developed by researchers who have been focusing exclusively on the benefits for their customers. Despite this, or precisely because of it, companies are now coming up against capacity and complexity limits. To enable further growth, they face the problem of having to optimize their production and logistics processes for efficiency.

TMC has specialized in supporting companies in this change process as an independent external partner. Such change processes start with an extensive analysis: How often do raw materials occur in existing formulations, how complex are formulations, and where are the bottlenecks in the processes? We identify similar raw materials to reduce their number. Based on the results of our in-depth formulation analyses, we identify consolidation and modularization potential. This potential is then proven by feasibility studies in our laboratory, where we develop optimized semi-finished products. Raw material consolidation is realized in close cooperation with R&D, production, and purchasing. In doing so, we always ensure the unchanged quality of the end products.

We pay particular attention to the optimization of production processes. The customer decides which suppliers they want to work with and to what extent they want to invest, if necessary. Even with minor investments, significant upturns in the capacity and effectiveness can be achieved. We closely track the market with our combined experience and network while maintaining our independence on suppliers.

TMC has established its own modern and high-performance laboratory capacities to demonstrate the optimization of formulations and processes practically. We thus support our customers who always have to manage change processes alongside their ongoing day-to-day business.

This lecture will present the complete analysis process and methods by which formulation and process optimization is achieved. These optimizations result in faster processes, better (more stable) quality and thus less rework, lower general process costs. Real-world scenarios will showcase the effects of such optimizations on the production capacity and process costs.

Mineral and Organic Fillers for Water-Dispersion Paints

Taras Karavayev [1], Olena Sevastyanova [2]

[1] State University of Trade and Economics (SUTE), Kyiv, Ukraine; e-mail: t.karavayev@gmail.com
[2] KTH Royal Institute of Technology (KTH), Stockholm, Sweden; e-mail: olena@kth.se

Water-dispersion (water-borne) paints (WDPs) are increasingly spreading in the world and Ukraine because of harmful solvents absence, ease application of coatings and its high performance properties and many other advantages compare with solvent based ones [1].
Mineral fillers (extenders) have significant influence on WDPs and coatings quality formation. It can partially replace expensive pigments improve the paints and coatings properties, perform specific functions: to regulate the paints rheological properties, be as reinforcing elements in the coatings etc.
According to our research [2] among carbonate fillers as the most convenient among chalks have been selected: fine natural chalk produced by PSC "Novgorod-Siversk plant of building materials" brand MMC-1; LLC "Sloviansk chalk-lime factory" brand MMC-2; LLC "Sloviansk industrial union "Soda" brand carbonate filler for plastics (CFP) and chemically precipitated chalk produced by LLC "Reaktyv". Among silicate fillers selected enriched kaolin from Prosyana deposit brands KC-1, KHF-86 and KBF-90 [3].
Comprehensive research of physical and chemical properties and structure of chalks and kaolins of main Ukrainian deposits have been conducted. Requirements to technical parameters of mineral fillers of WDPs have been created. The average particles size (0,7-5 micrometers), whiteness (73-95%), oil absorption value (20-58 g/100g) and other was defined as basic parameters. Mentioned proposed to supplement by the surface energy parameters: lyophilic ratio (0,1-1,25) and conventional dielectric loss tangent (0,02-0,20).
The quality of WDPs achieves through performance properties of obtained coatings. The most important factors of high-quality coatings is the creation of maximally packing density of the filler particles and their effectively combining with a binder. For this purpose, we proposed to use modifiers such as dispersant (for example sodium salt of polyacrylic acid – Axilat 32S) and repellent (potassium methyl silicon – repellent organic-silicon liquid (ROSL-11k)). The maximum useful effect is achieved by simultaneous application of both modifiers.
The composition of WDPs of different designation purposes created on the base of Ukrainian chalks and kaolins as mineral fillers that were selected as the result of their complex examination [4]. The results of our previous research have shown that Ukrainian chalks and kaolins are perspective mineral fillers of WDPs.
As fillers of WDPs can be used organic substances, for example lignin. This possibility was study during SUTE and KTH joint project realisation "Innovative composite materials based on the biopolymers and mineral fillers", which was financed by Swedish foundation STINT. Different types of technical lignin were obtained by KTH's researches. The small amount of lignin can be used as a filler, dispersant and surfactant in WDPs composition. This gives an opportunity to discuss the perspectives of using lignin as a functional additive in WDPs composition.

References:

[1] Karavayev T., Osyka V., Kolomiets T. Water-Borne Coating Materials Already Dominate. *European Coatings Journal*. V. 2, 18–20 (2019).
[2] Sviderskyi V., Karavayev T. Dispersion and structure of the carbonate fillers for water-dispersion paints. *Journal of Cherkasy State Technological University*, 2, 102–108 (2012).
[3] Sviderskyi V., Karavayev T. Composition and Physico-Chemical Properties of Ukrainian Kaolins Surface. *Chemistry and Chemical Technology*. V. 7, No. 2, 197–203 (2013) (in Ukrainian).
[4] Karavayev T., Sviderskyi V. (2012). Strength of films from water-dispersion paints filled with carbonates and kaolins. *Commodities and Markets*, 2, 180–190 (2012) (in Ukrainian).

UV curable biobased powder coatings

Dominika Czachor-Jadacka, B. Pilch-Pitera, B. Mossety-Leszczak, M.Kisiel

Rzeszow University od Technology, Rzeszow, Poland, d440@stud.prz.edu.pl

Powder coatings systems are becoming more and more popular due to their unique properties. Nowadays, the ecological aspect is important, and powder systems meet the following requirements. Due to strict environmental standards as well as care for the environment, UV curable powder system replaces solvent system because, on the contrary, they limit the release of volatile organic compounds (VOCs) [1]. From the application point of view, UV curing systems are primarily used in those applications where thermal curing is almost impossible, such as the curing of coatings on temperature sensitive substrates e.g. wood, paper and plastics. Ultraviolet (UV) curing has become a fast and environmentally friendly curing technique with many advantages such as high curing rate and good mechanical properties of crosslinked coating [2,3].

The main aim of the research was to obtain UV curable coatings intended for painting heat-sensitive surfaces. The study included the development of new bio-based crosslinking agents and optimisation of the powder coating composition. The synthesized crosslinking agents were based on diisocyanates and natural raw materials as well as acrylates containing hydroxyl groups. Among others, MDI, IPDI and TDI were used as diisocyanates and the natural components used are available polyhydroxyl alcohols (otherwise sugars) containing one hydroxyl group -OH at each carbon atom i.e. xylitol, mannitol, sorbitol, erythritol. The reaction of diisocyanates with sugar alcohols was catalysed with tin compounds. The obtained urethane crosslinking agents were used to prepare powder coating compositions. Apart from the obtained crosslinking agents, the powder compositions included unsaturated polyester resin synthesized by R&D Ciech Sarzyna SA, a flow control agent and a degassing agent - benzoine. Irgacure 651 and Irgacure 907 were used as the initiator.

The chemical structure of the urethane acrylates was verified by IR spectroscopy. UV-curable powder clear coatings were obtained through a combination of urethane acrylates with unsaturated polyester resins. The thermal behavior and crosslinking density were examined using DMA. The obtained coatings were evaluated by performing the following tests: roughness, gloss, scratch resistance, hardness, adhesion to steel and water contact angle. As part of this research, high hydrophobicity and scratch resistance of UV-curable powder clear coatings were developed, which are a VOC-free and economically attractive alternative method for low thermal resistance surface protection, such as for composites, wood and wood-based materials.

References:

[1] A. Srivastava, D. Agarwal, S. Mistry, J. Singh, UV curable polyurethane acrylate coatings for metal surfaces, Pigment & Resin Technology. 37, 217–223, (2008)

[2] R.S. Rawat, N. Chouhan, M. Talwar, R.K. Diwan, A.K. Tyagi, UV coatings for wooden surfaces, Progress in Organic Coatings. 135, 490–495 (2019)

[3] A. Javadi, H.S. Mehr, M. Sobani, M.D. Soucek, Cure-on-command technology: A review of the current state of the art, Progress in Organic Coatings. 100, 2–31 (2016)

Speed up your production
– Easy-to-disperse silica reduce the manufacturing time dramatically

M. Cornelius, M. Heuer

Evonik Operations GmbH, Essen, Germany,
e-mail: maximilian.cornelius@evonik.com

Fumed Silicas are known for more than 40 years as excellent rheological additives for paints, coatings, inks, adhesives etc.
Just by incorporation via a stirring or a milling process these additives will work properly, instandly and depending on the amount and the system as anti-settling agent, anti-agging additive or thixotropic modifier.
Next to these property improvements also the usage of a higher scratch and abrasion resistance in paints, inks and coatings is reachable by using special fumed silica products.
Without any influence to gloss, transparency, anti-corrosive behaviour, and chemicals resistance fumed silicas are the first choice to modify paints, coatings and inks.
The process of incorporation into a coatings resin matrix of the fumed silica is related to the kind of stirring process.
For this incorporation process a lot of shear forces are necessary to gain a good result of properly distributed fumed silica particles in the liquid coating for a good rheological result and as well an excellent coating surface without any specks or spots of large agglomerates respective particles.
So for primers and Tie-coats a longer treatment with a dissolver is sufficient while OEM clearcoats require a ball mill or a pearl mill incorporation.
The new developed easy-to-disperse fumed silicas are filling the gap for a faster production by a dramatic reduction of the incorporation time without impairing the rheological performance.
With these new easy-to-disperse fumed silica also the other chemical and physical properties for all kind of paints, coatings and inks will stay remained.
With lower amount of shear-forces also the option for a post-addition in ready-made coatings formulation is much more easier.
With these new products based on novel structure-modified fumed silica it is possible to give a contribution to the goals of the United Nations for more resource-efficient production of sustainable coatings.

New biobased polyurethane powder coatings for painting low temperature resistance substrates

B. Pilch-Pitera, D. Czachor-Jadacka, K. Pojnar, M. Kisiel, B. Mossety-Leszczak, M. Walczak

Rzeszow University od Technology, Rzeszow, Poland, barbpi@prz.edu.pl

An important direction in the development of the powder coating sector is the search for new powder coatings suitable for painting heat-sensitive surfaces. Standard powder coatings are cured at temperature above 170°C. This temperature restricts the range of their application for substrates with high thermal resistance, for example metal, ceramic or glass. The development of low-temperature powder products significantly expands their area of application for for materials with lower thermal resistance such as: MDF, wood and plastics.

The curing temperature of polyurethane powder coatings, depend on the deblockig temperature of polyisocyanate, which is a crosslinking agent in these type of products [1]. Blocking of isocyanate groups prevents the reaction between −NCO and -OH groups during extrusion and storage of powder coating [2,3]. The use of blocking agents that deblock at lower temperature leads to low temperature curing polyurethane powder coatings.

The main aim of this work was to develop of the new blocked polyisocyanates based on renewable substrates which were used as crosslinkers for low temperature curing polyurethane powder coatings. The new blocked polyisocyanates were based on diisocyanates and renewable materials as well as blocking agents containing hydroxyl groups. IPDI and TDI were used as diisocyanates and renewable components used are available polyhydroxy alcohols (otherwise sugars) containing one hydroxyl group -OH at each carbon atom i.e. xylitol, sorbitol, erythritol. The reaction of diisocyanates with sugar alcohols was catalysed with tin compounds. The obtained crosslinking agents were used to prepare powder coating compositions. Apart from the obtained crosslinking agents, the powder compositions included polyester resin based on isophthalic acid and neo-pentyl glycol, acid value: 4–6 mg KOH/g, hydroxyl value: 30–45 mg KOH/g, a flow control agent and degassing agent - benzoin.

In order to confirm the chemical structure of obtained blocked polyisocyanates, the Fourier transform infrared spectroscopy (FTIR) and nuclear magnetic resonance spectroscopy (NMR) were used. The molecular weight distributions were investigated using gel permeation chromatography (GPC). The process of deblocking of the obtained polyisocyanates were examined using DSC. The powder coatings were manufactured by means of extrusion process at 90°C and then cured in 145°C. Several properties of cured coatings were investigated such as roughness, gloss, thickness, adhesion to the surface, cupping and hardness.

The work resulted in new polyurethane powder coatings that may be promising opportunity for liquid paint for coating heat-sensitive substrates. The use of polyhydroxy alcohols results in higher functionality and higher molecular weight of polyisocyanates. Higher functionality of polyisocyanate leads to higher crosslinking density of powder coatings and better application properties. In addition, by using appropriate blocking agents, the deblocking and crosslinking temperature has been reduced.

References:

[1] Greszta-Franz D., Itter U., Crosslinking behavior of one component polyurethane coatings based on malonate blocked polyisocyanates: The effect of the malonate structure and optimization of the NCO:OH ratio. *Farby i lakiery*. 6, 3-8, (2012)
[2] Pilch-Pitera B., Byczyński Ł., Study on the thermal behavior of new blocked polyisocyanates for polyurethane powder coatings. *Progress in Organic Coatings*. 101, 240–244, (2016)
[3] Wicks D.A., Wicks Z.W., Blocked isocyanates III: Part B: Uses and applications of blocked isocyanates. *Progress in Organic Coatings*. 41, 1–83, (2001)

Two Approaches for Achieving Improved Sustainability of Architectural Coating Binders

Dr. C. Tiemeyer, Dr. C. Cheikh

Wacker Chemie AG, Burghausen, Germany, constantin.tiemeyer@wacker.com

Key words: vinyl acetate-ethylene dispersions; architectural coatings; biomass balance; C^{14} content;

Concerns about the environmental impact of industrial products and their processes have been constantly growing over the last decade. Since carbon dioxide is a proven greenhouse gas, its emission and accumulation in the atmosphere is directly linked to climate change. In line with the United Nations' Sustainable Development Goal (SDG) 13, society needs to "take urgent action to combat climate change and its impacts."

While the whole economy needs to be decarbonized, the chemical industry must move away from fossil raw materials while keeping an eye on economic considerations. Understanding petrochemical value chains is key to unlocking potential solutions. This can be accomplished in a first, straightforward step without re-inventing chemistry, by preparing standard chemical building blocks via transformation into biomass sources [1]. For the VAE (vinyl acetate-ethylene) value chain, bio-based alternatives, such as bio-based acetic acid, would all be suitable technical drop-in alternatives. However, those solutions must additionally be widely available and affordable in order to contribute to a fundamental shift toward renewable feedstocks with a positive impact on the environment.

Today, WACKER is proposing two approaches based on different bio-based raw materials suitable for architectural coating binders. First, bio-based acetic acid can be produced from a by-product of the wood industry. The wood is sourced from sustainably managed forests and the resulting bio-based acetic acid is certified by PEFC, the Programme for the Endorsement of Forest Certification Schemes. This bio-based acetic acid has the same chemical properties as its fossil counterpart. Therefore, it enables WACKER to use this raw material as a direct drop-in for the standard reaction processes. The final products are made in accordance with the third-party certified biomass balance approach and show identical technical performance when compared to their equivalents based on fossil raw materials. Therefore, this method makes it possible not only to reduce the overall carbon footprint of a product but also to avoid specific investment costs for both binder and paint producers, who are transitioning toward an industry that consumes fewer fossil resources.

The second approach is based on modifying a standard vinyl acetate-ethylene dispersion with starch to create a C^{14}-detectable raw material. The selected starch is sourced as a side-product from the agrochemical industry. It is recovered from wastewater and treated to improve its chemical properties to make it suitable for architectural coating applications.

In summary, the two approaches share the same goal, i.e. the reduction of the use of fossil resources. The biomass balance approach is a direct drop-in solution that avoids investment in dedicated infrastructures and additional R&D development, which leads to a decrease in overall invest cost [2], while the incorporation of starch enables the formulation of C^{14}-detectable bio-based architectural coating binders.

References:

[1] Bio-Based Chemicals, a 2020 Update, IEA Bioenergy, 2020
[2] C. Cheikh, S. Baueregger, Renewable Resources in VAE-based coatings. *Eur. Coat. J.* (6), page 38, (2021)

Enhancing sustainability in the manufacture and disposal of paints and coatings

B. Naden

PRA World, Melton Mowbray, (UK), b.naden@pra-world.com

Improvements in the sustainability of the coatings industry has been on the agenda for a number of years, particularly in the reduction of volatile organic compounds (VOCs). Environmental and Health legislation regulating the levels of VOCs in coatings and longer-term emissions of other volatiles has driven a major shift in the coatings industry so that more than 80% of decorative paints now sold in the UK and Europe are water-based. The focus for the industry has now expanded to include energy and resource conservation, waste minimisation, improved process efficiency, and the use of renewable and recycled materials for enhanced sustainability. Approximately 10% of the environmental footprint of coatings is formulation-related, with the rest shared equally between that generated through raw material manufacture, and through downstream use and disposal.

The most in-demand pigment used in paints and coatings is titanium dioxide (TiO_2), found in over 90% of products because of the opacity, aesthetics, durability of colour and protection properties it provides. Production of TiO_2 involves high embodied energy, emissions of polluting waste streams, and the use of raw materials from scarce resources. Any improvements to the environmental footprint of TiO_2 use is to be welcomed, and recycling of pigment from the approximately 55 million litres of waste paint generated in the UK each year, much of which goes to landfill, is one opportunity to achieve this. This will have the additional benefit of helping to divert a problematic waste material from the rising plastic waste that is currently attracting a great deal of attention and concern.

Carbon black is the most common black pigment used for paints; specialty carbon black applications, which include coatings, polymers and printing inks, represent around 1MT of a market dominated by application in vehicle tyres. End-of-Life Tyres (ELTs) presents one of the most problematic sources of solid waste, due to the large volume produced and their inherent durability. Pyrolysis has potential to reclaim valuable materials and prevent harmful emissions but is still largely commercially unattractive. One of the most valuable pyrolysis products is carbon black which, when reclaimed, purified and reused, can contribute significantly to a circular economy.

PRA is partnered with manufacturers and research organisations in projects that are part-funded by the UK Government aimed at developing sustainable manufacturing in the paint and coatings industry. This involves the recovery of pigments by pyrolysis of waste tyres (carbon black) and waste waterborne paint (TiO_2), including the development of technologies and processes for separation of materials of interest from complex mixtures of waste materials, purification (and any surface functionalisation), and application of recovered material.

This has the potential to address the significant environmental burden of end of life tyres and waste waterborne paint, as well as providing a sustainable approach to the manufacture of pigments, contributing to the circular economy. In addition to application in paints and coatings, these recovery processes are being investigated to provide plastics additives and for additives in thermal insulation manufacturing.

Thermally conductive powder coatings

Markus Cudazzo, Cathleen Joachimi

Fraunhofer IPA, Stuttgart, (Germany), Markus.Cudazzo@IPA.Fraunhofer.de

Powder coatings, like almost all coatings, are thermal insulators. This leads to an inhomogeneous heat flow within the powder coating layer and through the substrate in case of temperature differences due to a (forced) temperature treatment process or in field use.

As an experimental setup, carbon-based materials or other inorganic materials were blended into powder-coatings, either as a dry-blend-mixture or as coextruded samples. The layer thicknesses and temperature treatment parameters were varied.

Common thermal conductivity measurements with laser-flash analysis, in which a short heat pulse generated by a xenon flash light uses a small temperature difference between the top and bottom of the sample as a measure of thermal conductivity, are possible in principle, but very time-consuming. The sample thickness of the pill to be produced is not comparable to the thickness of a powder coating layer. It is impossible to measure powder bulk with this method, as the scattered light passing through the sample falsifies the measurement result.
The Coatmaster® measuring method, well known for measuring the thickness of powder coating layers, especially in the non-temperature treated state, calculates the layer thickness by means of the thermal conductivity of the layer. A xenon flash light is used, but in the top-light method without the need of a liquid nitrogen cooling. The layer thickness must be known. Then, conversely, after modification of the measuring technique, it is possible to calculate the thermal conductivity of the powder coating layer without time-consuming sample preparation, directly on the powder-coated substrate in the non-temperature-treated or temperature-treated state.

For the first time in the history of electrostatic powder coatings it is possible, to develop more detailed models for understanding the complex thermal treatment process of sintering, melting and crosslinking of a powder layer. Measurements without the addition of a thermal conductivity agent show a significant difference in the thermal conductivity of the powder bulk compared to the melted/crosslinked state. The thermal diffusivity in the bulk material is $\approx 10^{-2}$ mm²/s and rises to $\approx 10^{-1}$ mm²/s during melting/crosslinking.

A significant increase in the thermal diffusivity in the powder coating layer can only be achieved if large quantities, e.g. 10 wt.% of conductive carbon black, are added to the powder coating, whereas for electrically conductive powder coatings, about 1 wt.% would already be sufficient. With the high quantities added, the thermal diffusivity of bulk material is 10^{-1} mm²/s and increases to 10^{0} mm²/s during melting/crosslinking. With the not so strongly color-giving thermal conductivity agents, as e.g. boron nitride, likewise a noticeable increase of the thermal conductivity without significant increase of the electrical conductivity can be obtained. This really simplifies the electrostatic application of this material.

The embedding of such high quantities of thermal conductivity agent is a challenge regarding production and processing. The field of application just for forced heat treatment is too sophisticated. Future applications in the field are rather in the area of thermal interface materials (TIMs) like thermally conductive adhesives, powder coatings for e-mobility components, coatings of components for heat recovery, etc.

References:

[1] Cudazzo, M.; Ondratschek, D.; Strohbeck, U.: *„Hocheffiziente Verfahren zur Pulverbeschichtung von Holzwerkstoffen"*, Programm Lebensgrundlage Umwelt und ihre Sicherung (BWPLUS), Förderkennzeichen: BWT 22003, Institut für Industrielle Fertigung und Fabrikbetrieb der Universität Stuttgart, 2005, S. 1 - 45

[2] Strohbeck, U.; Cudazzo, M.; Vogelsang, H.; Gedan-Smolka, M.; Tuschla, M.; Demmler, M.: *"Innovative technology for highly productive powder coating of coils and blanks."* European Coatings Congress. - Nürnberg, 28.03.2011 – 30.03.2011. - 14.4, 8 Seiten.

[3] Cudazzo, M.; Strohbeck, U.; Pulli, K.: *„Energie- und materialeffiziente Technologien der Pulver-beschichtung: Kompakte pistolenlose Pulverbeschichtung-Precoating-Schnelle Aushärtungsverfahren-Dust on Dust-Technologie"*, in: Stuttgarter Produktionsakademie: *Pulverbeschichtung heute und morgen : Das volle Potenzial der Pulverbeschichtung ausschöpfen*. Seminar, 8. November 2017, Stuttgart, 2017.

The effect of n-Hexyl glycol and pigment to binder ratio on deposition behavior and edge coverage of water borne epoxy urethane electrocoatings

Mehdi Haghi[1], *Zahra Ranjbar*[1,*]

1 *Department of Surface coatings and corrosion, Institute for Color Science and Technology, P.O. Box: 16765-654, Tehran, Iran*

Edge coverage of cathodic electrocoating has a significant effect on corrosion resistance of these kind of coatings.

The effect of hexyl glycol as a rheology modifier of electrodeposited film and pigment to binder(P/B) ratio in a electrocoating bath have been investigated at constant voltage deposition of cathodic amine modified epoxy urethane electrocoating.

Various ratio of P/B in an electrodeposition bath in present and absence of hexyl glycol has been studied. Dual wave scan and gloss measurement were used after applying white automotive top coat directly on CED to analyze appearance of final film.

The edge corrosion performance on the knife plate is evaluated by salt spray and EIS measurements.

Crass cut test was used to investigates on adhesion of electrocoating on phosphated steel and inter coat adhesion between electrocoat and top coat.

The study revealed that P/B ratio has pronounced effect on appearance and corrosion properties and also in presence of 0.15 % hexyl glycol edge coverage and appearance significantly were improved.

Digitalization of formulation developments via predictive sciences

Sander van Loon, Gwenola Le Mouee, Alice Lorgue, Jose Ignacio Martinez Sanchez;

VLCI, Amsterdam, Netherlands

Van Loon Chemical Innovations B.V. (VLCI, www.vlci.biz), Amsterdam, The Netherlands, s.vanloon@vlci.biz

The applied predictive formulation sciences, Hansen Solubility Parameters (HSP) and Hydrophilic Lipophilic Difference – Net Average Curvature (HLD-NAC) are very powerful to find matching ingredients, resulting in improved stability and efficacy of end-products. These models are applicable to solutions, dispersions and emulsions, which basically includes all types of formulated products. Although it has been applied for many years, there is still limited use of them in coating formulation developments and ingredients thereof. The equations require practical parameters of the ingredients. Once generated, compatible combinations can be predicted to develop and optimize specific formulations. The ingredient parameters generated via the models are predictive and sustainable: you can use them over and over, allowing you to move away from trial-and-error and use digitalization in product developments. This is a very efficient way to enhance the properties, in addition to reducing the complexity, time and cost of developing coating formulations or ingredients. When combined with High Throughput (HT) screening for automated, parallel and small-scale preparation of samples and end-products, further efficiencies can be achieved.

The HSP and HLD-NAC approach and the required ingredient parameters will be explained via practical applications, to showcase how it can lead to efficient developments of a broad range of coatings and ingredients. The use of HT screening will also be explained and why this is needed to fill up the ingredient database of our predictive formulation apps. These apps can be used as the first stage of experimentation by formulating products digitally, followed by a drastically streamlined amount of practical lab work needed, compared to trial-and-error. This digital formulating approach will be demonstrated, to show how this can boost efficient coating and ingredient developments.

Submission for oral presentation at ETCC congress on 2-4 September 2020, Krakow, Poland.

Topic: ***Advances in Sciences of Paints & Coatings, Adhesives, Construction Materials, Printing Inks and related products***

From the synthesis of new biphenyl photoinitiators to the formation of highly cross-linked polymer products for 3D printing and coating industry

Wiktoria Tomal, e-mail: wiktoriatomall@gmail.com

Cracow University of Technology, Faculty of Chemical Engineering and Technology, Warszawska 24, 31 155 Cracow, Poland

Nowadays, polymerization processes are one of the most widely used in chemical industry. One of the recent and rapidly developing methods of obtaining polymers materials is the polymerization reaction initiated by light, i.e. photopolymerization. Photopolymerization has many application advantages, thanks to which it is widely used, among others, in the polymer coating industry, photolithography, medicine, microelectronics and 3D printing. The popularity of photopolymerization in so many areas is primarily due to its high speed, which can occur even in fractions of seconds. Moreover, the possibility of performing photopolymerization processes at ambient temperature makes the production of polymeric materials one of the most efficient photochemical technologies.

The multitude of applications of light-initiated polymerization processes has resulted in increased interest of scientists and companies in searching for new types of high-performance photoinitiators, as it is their properties that determine the efficiency and speed of polymerization. In addition, light sources such as UV-LED and Vis-LED are starting to be used on the photochemistry processing market as an environment-friendly alternative to the previously used medium pressure mercury lamp (MPM). The problem, however, is the lack of compatibility of absorption characteristics of commercially used photoinitiators with the characteristics of emissions of industrial light sources (UV-LEDs, Vis LEDs).

Novel biphenyl derivatives were proposed as photosensitizes of iodonium salt for a highly effective bimolecular photoinitiating system upon soft irradiation conditions under long-wave ultraviolet (UV-A) and visible light (Vis). Remarkably, these structures are highly versatile allowing access to photoinitiating systems for the free-radical polymerization of acrylates, the cationic photopolymerization of epoxides, glycidyl and vinyl ethers. Furthermore, they are also efficient in the process of obtaining interpenetrated polymer networks (IPNs).

Acknowledgements:

This work was supported by the Foundation for Polish Science within the project TEAM-TECH (Contract No. POIR.04.04.00-00-204B/16-00 - TEAM TECH/2016-2/15 – “Molecular design, synthesis and application of photoinitiator-catalysts (PICs) for photopolymerization reactions”).

The author is also grateful to the project The project "WAY TO EXCELLENCE - a comprehensive university support program" implemented under the Operational Program Knowledge Education Development 2014-2020 co-financed by the European Social Fund WND-POWR.03.05.00-00-Z214 /18

New approach of hybrid layer on titanium dental implants

Alicja Kazek-Kęsik[1], *Katarzyna Leśniak-Ziółkowska*[1], *Monika Śmiga-Matuszowicz*[1], *Monika Brzychczy-Włoch*[2], *Wojciech Simka*[1]

[1] *Silesian University of Technology, Gliwice, Poland*
[2] *Jagiellonian University Medical College, Cracow, Poland*

Bone implants are usually made by titanium or medical stainless steel [1]. For these materials, the surface treatment is provided to enhance their biocompatibility and bioactivity properties. Titanium (Ti Grade 4) and its alloy Ti-6Al-4V is biocompatible, but not exhibit antibacterial properties. One reason of implant failure is septic inflammation. To avoid this process after material implantation the patients take an antibiotic in their high concentration (0.25 – 1.0 g per day). According to the World Health Organization on Surveillance of Antibiotic Consumption report (2016-2018) prescriptions for antibiotics should be monitored to avoid formation a large group of new antimicrobial agents. Metabolic products of the antibiotics also significant influence on our environments. When we take into consideration that number of implantation surgeons still increase, the amount of antibiotic taken by patients also increasing.

Titanium implants might be modified by two techniques such as plasma electrolytic oxidation (PEO) and dip coatings. PEO process allows to modify surface of implants with various shapes, and porous oxide layer is easily form. The oxide layer is usually composed of bioactive compounds which strongly influences on osseointegration process [2]. Dip coating method allows to make a thin polymer layer with small amount of antibiotics to prevent bacteria adhesion. New approach of this layer is fast-degradation process of polymer and release antibiotic in their stable, active form. It is possible to form the polymer layer with concentration of antibiotics up to 100 $\mu g/cm^2$ of implant surface. For example, surface of β-phase titanium alloy (Ti-15Mo) was modified by PEO process in solution composed of $Ca(H_2PO_2)_2$. Then, fast-degradable polymer poly(sebacic anhydride) was deposited on the top of the surface with amoxicillin in concertation of 96.45 μg per implant sample. It was determined that after 0.5 h of coating immersion, 20.85% of amoxicillin was released into artificial saliva. It was enough to inhibit growing of *Staphylococcus aureus (DSM 24167) and Staphylococcus epidermidis (ATCC 700296)* bacteria. Hybrid coating was degraded in more that 77 mol.% after 7 days of the titanium samples immersion. This surface was also cytocompatible with osteoblast-like MG-63 cells after 7 days of culture.

New approach of this work is formation the implants with the desirable properties, using the small amount of biologically active substance and release it in short time. Formation of additional polymer layer do not significant influence on the porous layer morphology which is favorable for formation bone tissue.

[1] Cordeiro J.M., Beline T., Lúcia A. et al. Development of binary and ternary titanium alloys for dental implants. Dental Materials. 33(11), 1244-1257, (2017).

[2] Krząkała A., Kazek-Kęsik A., Simka W., Application of plasma electrolytic oxidation to bioactive surface formation on titanium and its alloys. RSC Advance. 3, 19725-19743, (2013).

Interaction and diffusion phenomena in 2K water-borne polyurethane systems for industrial coatings

S. Kirsch, Q. Kainz, F. Lucas, S. Roller,
BASF SE, Ludwigshafen, Germany
M. Soleimani,
BASF Corp., Wyandotte MI, USA
Y. Liu, H. Zhou, K. Tran, J. Schackmann, M. Zhang, M. Winnik,
University Toronto, Canada

e-mail to: stefan.kirsch@basf.com

Due to environmental concerns, lowering VOC in modern, industrially applied coatings is one of the major driving forces for innovation in various applications. Two strategies can be observed in the market place: first, reducing solvent leading to ultra-high solid formulations still having an adequate viscosity range and second, to convert 2K PU to water-based systems. The challenge of the latter is to cope with a significantly different process of film formation compared to solvent-based systems. This is due to the heterogeneous existence of both, the binder and the polyisocyanate, dispersed in the water-phase. Additionally, the polyisocyanate hardener can undergo a wide variety of reactions with water as well as with the binder, competing each other. This can tremendously impact both optical and mechanical properties of the resulting films.

In this paper we investigated the diffusion of the key components based on aliphatic polyisocyanates in the system during different states in the process of film formation. This is the reason why a plethora of analytical methods have been tested and developed in order to deepen the existing knowledge. In the liquid state analytical ultracentrifugation (AUC) has been used to characterize the interaction of polyisocyanate droplets with the surrounding dispersion particles. The impact of each component has been consequently studied. In the critical phase of film formation, where the volume fraction of polymer phase rapidly increases, diffusion and dissolving of those droplets is monitored by means of confocal laser scanning microscopy (CLSM). In the final stage of film formation, the ability of the individual polymer chain to diffuse to the reactive NCO sites creating a dense network is crucial. Here, fluorescence resonance energy transfer (FRET) has been applied to study the polymer diffusion in the latex film. The know-how gained is translated into new product range of binders (Basonol® AC) and hardeners (Basonat® HW) for water-based 2K PU applications.

Production of polymer coatings by photopolymerisation using novel, effective photoinitiators and simultaneously colorimetric sensors to control the quality of the process

Monika Topa

Faculty of Chemical Engineering and Technology, Cracow University of Technology, Cracow, (Poland), e-mail: monika.topa@doktorant.pk.edu.pl

Photoinduced polymerisation of monomers is gaining popularity because it is an environmentally friendly and safe method for production of protective polymer coatings. Moreover, it has plenty of benefits. The basic advantage of the photopolymerisation over other methods used for production of polymer coatings is its speed. These compositions are transformed from a liquid state into a fully cross-linked solid within seconds without emission of any volatile solvents to atmosphere. The photopolymerisation has other advantages, e.g.: ambient temperature operation, solvent free formulations, and low energy consumption [1,2].
Because of many advantages, photopolymerisation is used in many scientific disciplines. The most popular is the coating industry. By photopolymerisation it is possible to obtain polymer coatings for automotive and furniture industries.
Nevertheless, the quality of the obtained polymer coatings is important. During photopolymerisation, many physicochemical parameters change, i.e. viscosity, polarity, crosslinking density, etc. Accurate tracking of the kinetics of changes occurring during ultrafast reactions is a difficult challenge when high intensity radiation sources are additionally used in industrial practice [3]. For this purpose, e.g. colorimetric fluorescent probes, which are the subject of this work, can be used.

In this work, cationic monomers such each (3,4-epoxycyclohexane)methyl 3,4-epoxycyclohexylcarboxylate (CADE) and triethylene glycol divinyl ether (TEGDVE) were applied as an epoxy and vinyl model monomers for cationic photopolymerisation for production of coating materials. The progress of the fotopolimerisation of cationic monomers and novel compounds as colorimetric sensors and simultaneously photoinitiators was studied using real time-FT-IR method. The system was composed of the Thermo Scientific i10 Nicolet^{TM} spectrometer with an appropriate horizontal adapter adapted for on-line measurements of photopolymerisation processes.

Acknowledgments
This work was supported by the Ministry of Science and Higher Education within the project Diamentowy Grant (project no.0052/DIA/2018/47).

References:

[1] Wolfram, S. Polymers and Light, Fundamentals and Technical Applications. *WILEY-VCH Verlag GmbH & Co. KGaA, Weinheim.* (2007)
[2] Reinhold, S. UV coatings: basics, recent developments and new applications. *Elsevier.* (2007)
[3] Joanna, O., Mateusz, G., Roman R., Dariusz, B. Application of a carbazole derivative as a spectroscopic fluorescent probe for real time monitoring of cationic photopolymerization. *Polish Journal of Chemical Technology.* 16 (1), 75 – 80, (2014)

Acrylic Emulsion Binder Extending the Aesthetics & Durability of Façade Coatings

J. Vyörykkä[1], D. Ressnig,[1] J. Sobczak,[2] M. Westmeyer,[2]
1: Dow Deutschland Anlagengesellschaft mbH (Germany), e-mail: jvyorykka@dow.com
2: Dow (USA)

Exterior thermal insulation composite systems (ETICS) are multilayered systems that are installed to building façades to improve thermal insulation. ETICS are expected to have a long service-lifetime to compensate installation costs with gains through energy savings. The durability is strongly affected by defects introduced through unfavourable interaction with water and humidity (e.g. cracks, mold or algae growth). We present technical challenges and investigations that led to the development of a silane modified acrylic binder for extended durability in exterior coatings, specifically for ETICS.

The Façade protection theory outlined by Heilen, first postulated by Künzel in the 1960's, offers complex relationships regarding the durability of façades and their properties.[1-3] Water resistance cannot be achieved by preventing water from entering an organic façade, but rather through moisture management. Moisture management is defined as balancing the moisture resistance and the breathability of the façade. The hydrophobic capacity of the façade must be achieved through the reduction of capillary water absorption (w-value) without affecting water-vapour permeability (V or Sd value).

The binder used in the outer ETICS layer (plaster or paint) has a critical impact on the performance against these two properties. The combination of silicone resin with acrylic emulsion binders has continuously gained in popularity due to favorable moisture management characteristics, regarding W- and Sd values. In these dispersion-silicone resin coatings, parts of the acrylic binder is replaced with Si-resin to increase water vapor permeability at low water pick up. This adaptation however imposes other limitation to the plaster or paint formulation, such as long term colour retention. Colour fade is in many cases also a sign of binder degradation that ultimately results in cracking, chalking and loss of water barrier properties of the coating.

We present a study on the relation between color retention and water/moisture management in silicone plasters as a basis to develop a polymeric binder concept that addresses this limitation.

References

[1] Baeră, C.; Szilagyi, H.; Dico, C. "Moisture Transfer through Façades Covered With Organic Binder Renders", *Construcŧll*. 14 (1), 70-? (2013)
[2] Teleng-van Oorschot, H. "Colourful façade coatings with inorganic colorants", Architectural Coatings, PPCJ, March (2012).
[3] H. Kittel, Lehrbuch der Lacke und Beschichtungen, Band 3, S. Hirzel Verlag Stuttgart, Leibzig 2001, 265.

Toughen your Architectural formulations with innovative particles - Novel Synthetic Silica Engineered for Improved Durability and Burnish Resistance

Oliver Peters, Jonas Berg, Ellen Reuter

Evonik Operations GmbH, Goldschmidtstr. 100, 45127 Essen, (Germany),
e-mail: oliver.peters@evonik.com

One challenge in formulating Architectural coating systems is to combine good aesthetical properties with superior mechanical robustness.

Typically, in Europe, customers ask for a very good hiding power and rather low gloss for interior applications - which is achieved by formulating with high filler & pigment content, respectively. In addition to visual appearance, there are high expectations on mechanical robustness of the films e.g. regarding wet scrub resistance or burnish resistance – in other words the resistance to change in surface appearance when mechanical stress is applied. While wet scrub resistance is strongly influenced by binder choice, burnish resistance can be improved by choosing the right particles. Evonik has developed a new class of specialty silicas with an innovative particle morphology.

During the production process a unique spherical particle design is obtained with a very narrow particle size distribution. The particle size can be tailored so that different aspects can be addressed depending on the requirements.

Understanding the unique performance benefits these materials can impart helps to optimize formulations. Results from profilometric studies & scanning electron microscopy show significant differences to standard formulations when these innovative materials are used.

PhotoDSC study of Titanium Dioxide Pigmented UV-cured Coatings: Anatas Polymorph against Rutile Polymorph

Hossein Yahyaei, Tahere Rostami, Mohsen Mohseni, Atefe Asgharnia

Department of Polymer Engineering and Color Technology, Amirkabir University of Technology, P.O. Box 15875-4413, Tehran, Iran

Due to the limitative environmental regulations about the release of the VOC, using of eco-friendly coatings was extended. The radiation curable coatings which are cured by exposing to a radiation of ultraviolet (UV) or electron beam (EB) achieved an important contribution in extending of eco-friendly coatings. Conversion can impact final properties of coating and efforts are underway to improve final conversions in UV cure coatings. However, about pigmented coatings the most important problem is absorption and or scattering some portion of the ultraviolet emitted by pigment particles. Titanium dioxide is the most useable white pigment with high opacity and hiding power. High refractive index causes high scattering and absorption in UV could be the reasons of reducing conversion. The range of band gap for titanium dioxide polymorphs is 3.0 eV to 3.4 eV that is in accordance with 365 nm to 380 nm wavelengths. Therefore, titanium dioxide electrons can excite and absorb the radiated UV. Thus, pigments prevent the UV radiation to absorb by the photoinitiators. As was mentioned above, the curing of white pigmented UV cured coatings is a prime challenge and rutile and anatase titanium dioxide have different band gap, different adsorption and in consequence different behavior in UV curing coating. The aim of this work is investigations of the effect of two different crystal structures of titanium dioxide (rutile and anatase) on the final conversion and conversion rate of the pigmented coating and comparing with unpigmented coating. UV-Visible spectroscopy showed the more absorption of rutile polymorph than anatase in the spectrum range of 350 to 420 nm which photoinitiators usually activated in this region. PhotoDSC results confirmed that final conversion, rate of conversionfor anatase was more than rutile. Also, TPO photoinitiator showed better performance than ITX.

References:

[1] Farzad , H., Najafi, F. Synthesis and Characterization of Aliphatic Tri-functional Oligomeric Urethane Methacrylate used for UV-Curable Aluminum Pigmented Coatings. *Journal of Macromolecular Science*,. 50, 504 – 512, (2013)

A coating from nature

George Hermens, Thomas Freese, Keimpe van den Berg, Rogier van Gemert, Ben Feringa

For almost a century, petrochemical-based monomers like acrylates have been widely used as the basis for coatings, resins, and paints. The development of sustainable alternatives, integrating the principles of green chemistry in starting material, synthesis process, and product function, offers tremendous challenges for science and society. Here, we report on alkoxybutenolides as a bio-based alternative for acrylates and the formation of high-performance coatings. Starting from biomass-derived furfural and an environmentally benign photochemical conversion using visible light and oxygen in a flow reactor provides the alkoxybutenolide monomers. This is followed by radical (co)polymerization, which results in coatings with tunable properties for applications on distinct surfaces like glass or plastic. The performance is comparable to current petrochemical-derived industrial coatings.

A New Styrene Acrylic to Manage Formulation Costs While Maintaining High Performance

Jeaneta Zieglerova

Synthomer Deutschland GmbH, Marl, Germany,
e-mail: jeaneta.zieglerova@synthomer.com

In decorative wall paint formulations, it is a constant objective to optimise formulation costs while maintaining or even increasing end-use performance. In addition, Volatile Organic Compound (VOC) reduction trends lead formulators to select binders with low Minimum Film Formation Temperature (MFFT) that can impact paint durability.

A newly developed styrene acrylic waterborne binder from Synthomer can help formulators solving this challenge. Internal application results show how this new low MFFT binder enables an improvement in wet scrub resistance and pigment binding. This gives formulators better latitude to optimise ingredient levels either for managing costs or enhancing end-use performances. This will be demonstrated in matt and semi-gloss paint formulations.

Diethanolamine Borate Modifiers for Water-Dispersion Polymeric Materials

Yevhenii Levchenko, Oleh Chervakov, Olha Sverdlikovska, Denys Chervakov

SHEI "Ukrainian State University of Chemical Technology", Dnipro, Ukraine
e-mail: theleva7@gmail.com

Since the creation of the first synthetic binder-based paints and varnishes, the researchers have been faced with the task of creating coatings with improved mechanical and decorative properties. One of the main methods for imparting high decorative and protective properties to coatings is to increase the hardness of the coating film which is achieved, among other things, by creating new binders or additives that modify their properties.

In the case of waterborne coatings improvement of protective properties is achieved by use of polymer dispersions with minimum film forming temperatures (MFFT) above room temperature. To reduce the MFFT of aqueous dispersions, coalescents are used. In the past the number of organic compounds of differing composition, namely hydrocarbons, esters, ethers and other volatiles were offered to be used as coalescents. But due to the tightening regulations on the VOC content in waterborne coating, coalescents exhibiting low volatility were developed. One particularly well known example of these low volatility coalescents is the 2,2,4-trimethyl-1,3-pentanediol monoisobutyrate (Texanol®) with a boiling point above 254°C. At the same time, use of ionic compounds as coalescents in waterborne coatings remains underresearched.

Formation of cyclic diethanolamine borate ester (DEAB) by interaction of diethanolamine and boric acid has been previously described by Koteneva et al. [1].
Previously we have shown [2, 3] that in aqueous solutions products of said interaction can form ionogenic compounds capable of ion exchange. They have been used to synthesize a number of ionic compounds (were X^- – anion of organic or inorganic acid) with the general formula:

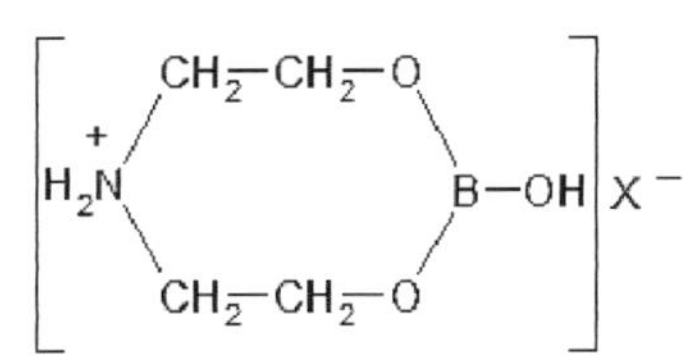

These compounds have shown similar performance to Texanol®.

Films formed by the paint based on styrene-acrylic dispersion Tritex SA-50 containing DEAB exhibited similar decorative properties to those with Texanol®. Replacing Texanol® with DEAB resulted in 17% higher film hardness.

Films obtained by drying urethane-acrylic latexes containing DEAB or Texanol® exhibited nearly identical mechanical and decorative properties.

References:

[1] Koteneva, I. V., Sidorov, V. I., Myasoyedov, Ye. M. O vzaimodeystvii bornoy kisloty s mono- i dietanolaminami. *Khimicheskaya promyshlennost' segodnya*. 8, 35–40 (2013) (In Russian).

[2] Sverdlikovska, O., Chervakov, O., Fedenko, O., Koshel, S., Levchenko, Ye. Polimerni ionni ridyny ta ionni ridyny ionenovogo typu: monograph. 147-160, (2020) (In Ukrainian).

[3] Levchenko, Y., Sverdlikovska, O., Chervakov, D., Chervakov, O. Development of coalescents for paints and varnishes based on ionic liquids – the products of diethanolamine and inorganic acids interaction. *Eastern-European Journal of Enterprise Technologies*. 2 (6 (110)), 21–29, (2021).

Synthesis of Rubber Seed Oil Waterborne Alkyd Resin from Glucitol

[1]Chigozie Francolins Uzoh

[1]Department of Chemical Engineering, Nnamdi Azikiwe University, Awka,

Faculty of Engineering, Nnamdi Azikiwe University, Awka.

*Corresponding author: E-mail: cf.uzoh@unizik.edu.ng

Phone: +2348033762029.

Abstract

Surface coating is a universal explanation for materials that may be applied as a covering or continuous film to the surface of an object (commonly refers to as the substrate) and dry to produce adhering films. Alkyd resins are major components of most surface coatings. Vicissitudes in consumer choice of surface coating and VOC regulation have constrained industries to decrease the VOC in coating. In most circumstances this has compelled a shift from solvent-based to water-based technology. Solubility properties of alkyd resin from phthalic and maleic anhydride, glucitol and rubber seed oil (RSO) as raw materials were examined. Four types of glucitol based medium oil length alkyd resins (Alkyd A-D, two for each acid anhydride) samples were formulated with phthalic anhydride, maleic anhydride, glucitol aqueous solution and RSO. The physical, chemical and film characteristics of the glucitol modified alkyd resins were examined and the result compared with standard alkyd resins and soybean based alkyd resin. It was observed that all samples were soluble in xylene and butoxyethanol but not water. The entire samples have varying solubility in water: xylene ratio (%) and complete solubility in water: butoxyethanol (%) ratio of 40:60, 50:50 and 60:40, respectively). ALKYD A is soluble in all the solvents except water where it was sparingly soluble at room temperature. The spectral analysis revealed the polyesterification reaction and hydrogen bonding integrity of the alkyd resins and the polybasic acid. The physical and chemical properties and the performance indices of the glucitol based alkyd resin results show that they possess acceptable drying properties, chemical resistance and mechanical properties. Performance of ALKYD A and C gave nearly the same result as the standard alkyd resins and soybean based alkyd resin. It has potential as a choice waterborne binder for alkyd resin paint and can help in reducing VOC.

Cataphoretic deposition of organic coatings with functionalized graphene oxide flakes

M. Calovi, S. Rossi, F. Deflorian, S. Dirè, R. Ceccato

Department of Industrial Engineering, University of Trento, Trento, (Italy), massimo.calovi@unitn.it

Graphene is a material object of studies in the field of composite materials, exploiting its excellent electrical [1,2], thermal [3,4] and mechanical properties [5]. In recent years, it has been attracting a strong interest in the organic coatings production, as it allows to create multifunctional layers [6]. Graphene flakes, in fact, can act as a barrier to the passage of aggressive ions, improving the corrosion protection of the coating [7], and at the same time increasing the conductivity of the organic layer. In this work the cataphoretic electrodeposition process was used for the realization of smart coatings, as this method guarantees to obtain layers with good adhesion properties, in a simple and economic way [8]. The graphene, as raw material, has been subjected to nitric acid oxidation treatment and functionalized with three different organosilane, to get a good dispersion in the acrylic cataphoretic bath. The samples' morphology was observed under optical and electronic microscope, to study the possible defectiveness of the coatings. The behavior in an aggressive environment was evaluated by exposing the samples in salt spray chambers and their corrosion resistance properties were verified through electrochemical impedance spectroscopy measurements. The process parameters, as well as the concentration of functionalized graphene oxide inside the cataphoretic bath, have been optimized, obtaining composite coatings with excellent corrosion resistance properties. It has been verified that the presence of graphene in the cataphoretic coating allows to obtain a better behavior in an aggressive environment and high guarantees from the point of view of the protection of the metallic substrate. The flakes enabled the reduction of the cathodic delamination and the increase of the abrasion resistance of the acrylic cataphoretic matrix.

References:

[1] Geim, A.K., Novoselov, K.S. The rise of graphene. *Nat. Mater.* 6, 183–191, (2007)

[2] Pokharel, P., Truong, Q.T., Lee, D.S. Multi-step microwave reduction of graphite oxide and its use in the formation of electrically conductive graphene/epoxy composites. *Compos. Part B Eng.* 64, 187–193, (2014)

[3] Tian, M., Qu, L., Zhang, X., Zhang, K., Zhu, S., Guo, X., Han, G., Tang, X., Sun, Y. Enhanced mechanical and thermal properties of regenerated cellulose/graphene composite fibers. *Carbohydr. Polym.* 111, 456–462, (2014)

[4] Shahil, K.M.F., Balandin, A.A. Thermal properties of graphene and multilayer graphene: Applications in thermal interface materials. *Solid State Commun.* 152, 1331–1340, (2012)

[5] Lee, C., Wei, X., Kysar, J.W., Hone, J. Measurement of the elastic properties and intrinsic strength of monolayer graphene. *Science*. 321 (80), 385–388, (2008)

[6] Atif, R., Shyha, I., Inam, F. Mechanical, thermal, and electrical properties of graphene-epoxy nanocomposites - A review. *Polymers.* 8, 281-318, (2016)

[7] Ramezanzadeh, B., Ghasemi, E., Mahdavian, M., Changizi, E., Mohamadzadeh Moghadam M.H. Covalently-grafted graphene oxide nanosheets to improve barrier and corrosion protection properties of polyurethane coatings. *Carbon.* 93, 555–573, (2015)

[8] Almeida, E., Alves, I., Brites, C., Fedrizzi, L. Cataphoretic and autophoretic automotive primers: A comparative study. *Prog. Org. Coatings.* 46, 8–20, (2003)

Modelling of Polymer and Solvent Parameters For The Prediction of Coating Surface Phenomena

Yannik Schuchmann, Jochen Stefan Gutmann, Christian Schmitz

University of Applied Sciences Niederrhein, Institute of Coatings and Surface Chemistry Krefeld, Germany, Yannik.schuchmann@hs-niederrhein.de

Statistical methods are strong tools to optimize chemical reactions or coating formulations. Conventional design of experiments in combination with polynominal linear regression can deliver information about the optimal composition of formulated materials such as coatings, but they require many samples, lack accuracy and grant no information about the chemical interactions between the different components of the coating system and thus do not match the current technical standards. Modern machine learning approaches like gaussian processes meet the requirements of Industry 4.0 and offer opportunities to acquire models. The aim of this work is to do a modelling for certain surface parameters of hardened coatings, which are known to be dependent on the polarity of the ingredients, especially resin and solvent polarity.

The properties of the coating can be related to the chemical structure of the ingredients by describing interactions with metrics of the polarity. In the coating industry the Hansen solubility parameters (HSP) is one well-known concept but suffers from poor measurability of these values of resins and polymers. Another approach are the $E_T(30)$ and Kamlet-Taft scales [1], which are accessible by the absorption maxima from probes being dissolved within the media. Moreover, the composition of the coating and these parameters of the ingredients can be used creating a predictive model applying machine learning techniques.

Different methacrylate resins were synthesized using three different monomers with different functionalities and polarities. These resins were characterized by molecular weight, molar ratio of the copolymer and its polarity parameters. One key methodology was incorporation of the probes and detection by UV-vis spectroscopy of the dried polymer films. Furthermore, acrylate/melamine formaldehyde formulations with various solvents were produced and coating properties like gloss, haze and surface defects were characterized. Finally, predictive models, so-called digital twins, were calculated by relating ratio of the ingredients and the aforementioned parameters to the coating properties. The validities of the models were verified by cross validation and information were derived by sensitivity analysis from the model.

The overall result will be a model which gives the opportunity to trace back surface phenomena which depend on polarity relations to certain molecular groups which are incorporated into the used macromolecules. These models will give further knowledge about the concrete interactions of functional groups and thereof the effects on the coating performance.

[1] S. Spange et. al., „Empirical polarity parameters for various macromolecular and related materials", Macromol. Rapid Commun. 21, 643–659 (2000)

Novel versatile bimolecular photoinitiating systems for coating industry and 3D printing application

E. Hola, J. Ortyl

Cracow University of Technology, Cracow, (Poland),
e-mail: emilia.hola@doktorant.pk.edu.pl

The most modern technologies for producing polymeric materials are based on photochemically initiated processes. The growing interest in photopolymerization prompts the search for new types of high-efficiency photoinitiating systems, because their properties determine the efficiency and speed of polymerization. Currently, it is very important that the new initiating systems allow operation using ultraviolet (UV) and visible (Vis) light in the field of emissions of UV-LEDs and Vis-LEDs.

Application of synthesized compounds for the role of photosensitizers in two-component photoinitiating systems (with iodonium salt - Bis-(4-t-butylphenyl)-Iodonium hexafluorophosphate or with amine - ethyl 4-(dimethylamino)benzoate) was investigated. Fourier transform infrared spectroscopy (real-time FT-IR) was used to monitor the kinetic of photopolymerization processes in the presence of studied photoinitiators.

Bimolecular initiating systems were used for different types of photopolymerization, cationic ring-opening photopolymerization of epoxides, free-radical photo-polymerization of acrylates, thiol–ene photopolymerization of allyl and thiol monomers or synthesis of interpenetrating polymer networks (IPNs) from acrylate/epoxy blends. As a light source Light Emitting Diodes (LEDs) with maximum emission at 365 nm, 405 nm, and 420 nm were used.

Moreover, two different initiation mechanisms were also analyzed for bimolecular photoinitiating systems with iodonium salt or amine through steady-state photolysis experiments, fluorescence quenching experiments, laser flash photolysis, theoretical calculations of molecular orbitals, and electrochemical analysis.

Studies have shown that bimolecular photoinitiating systems have extensive applications in different types of photopolymerization processes across the photo-oxidation or photo-reduction mechanism of their excited species promoted by soft UV-A and visible light. Investigated photoinitiating systems were also applied for photosensitive resins for laser writing/ 3D printing experiment upon exposure to laser diode 405nm.

Acknowledgments:

This work was supported by the Foundation for Polish Science (Warsaw, Poland) TEAM TECH project Grant No. TEAM TECH/2016-2/15 (POIR.04.04.00-00-204B/16-00) "Molecular design, synthesis and application of photoinitiator-catalysts (PICs) for photopolymerization reactions".

References:

[1] Hola E., Pilch. M., Galek M., Ortyl J., Y. New versatile bimolecular photoinitiating systems based on amino-m-terphenyl derivatives for cationic, free-radical and thiol–ene photopolymerization under low intensity UV-A and visible light sources. *Polymer Chemistry*. 11, 480 – 495, (2020)

[2] Hola E., Ortyl J., Jankowska M., Pilch M., Galek M., Morlet-Savary F., Graff B., Dietlin C., Lalevée J., New bimolecular photoinitiating systems based on terphenyl derivatives as highly efficient photosensitizers for 3D printing application. *Polymer Chemistry* 11, 922 – 935, (2020)

Learning from swarms: mini sensors for measuring the layer thickness in an industrial environment

Dipl.-Physiker Georg Nelke
OptiSense Gesellschaft für Optische Prozessmesstechnik mbH & Co. KG
Annabergstraße 120
45721 Haltern am See
GERMANY
info@welltext.de

In the future, coating thickness measurement in the industrial environment should be carried out with high speed and accuracy with the highest stability. Ideally, the test systems can also measure contact-free, non-destructively and in motion.

In this lecture, novel sensor concepts are presented that meet the requirements of comprehensive quality controls outlined above. The focus is on a RAIS (Redundant Array of Independent Sensors) sensor system in which a large number of miniaturized sensors work together as a single large whole. The failure of individual sensors has no effect. This pioneering solution offers significantly more comprehensive and therefore better measurement results: from the weight per unit area to wet and dry layer thickness to porosity. The sensor network also ensures that no data is lost and that the measuring station for the system is permanently available.

Furthermore, you will learn in the lecture how state-of-the-art sensors and data processing components interact to minimize calibration processes and to integrate intelligent, efficient IIOT sensors.

Colour durability of PUR topcoats – how to choose durable colour?

D. Wojda, I. Kunce, A. Królikowska

Road and Bridge Research Institute, Warsaw (Poland), dwojda@ibdim.edu.pl

In paint systems designed for use in highly corrosive environments, the topcoat is usually made of polyurethane binders. They are commonly used due to their good resistance to weather conditions and economic advantages. Apart from its protective barrier function, the topcoat is also responsible for the aesthetics of the object being protected against corrosion. Above all, the aesthetics of the coating are assessed in terms of gloss and colour. In the case of new generation anticorrosive liquid paints, there are no recommendations for colour durability, which could be helpful for designers. For this reason, the aim of the study was to determine the colour fastness of the topcoat exposed to UV radiation and analyze their degradation process.
The study exposed samples, in 11 different colours, to UV radiation and moisture and then measured the changes in colour, gloss and thickness. The colour change was expressed with ΔEab and $\Delta E00$. Although $\Delta E00$ is less popular than ΔEab, it describes a change in colour that better correlates with the change perceived by the human eye. Changes in the structure of polyurethane topcoat were recorded by Fourier Transformed Infrared Spectroscopy (FTIR).
The greatest changes in colour were found in RAL 8004 (brown), RAL 3002 (red) and RAL 1016 (light yellow) coatings regardless of the parameter expressing them. The values of ΔEab for most colours were almost twice as high as the change expressed by $\Delta E00$. Samples in RAL 1016 and 6018 (green) are characterized by the highest gloss variation and both exceed 80%. Coatings in brown and light yellow had the highest change in thickness. The analysis of the coatings with FTIR showed that the degradation of all the coatings takes place according to the photodegradation process described in the literaturę. Differences in peak intensity changes result from different pigmentation of the coatings.

Coatings Appearance Measurements – new developments and trends

T. Wieckowski, H. Folkerts, B. Bidzinska

EUROTOM PTH Sp. z o.o., BYK GARDNER GmbH, Warszawa, (Polska), eurotom@eurotom.pl

The appearance assessment is a sophisticated process, finally verified by client's visual judgment. The efforts to give an objective picture of the sample require complex instrumental approach. The presentation focuses on new techniques in gloss, colour and structure measurement needed to fulfil this task. Multifunctional devices are invented and introduced combining gloss, colour, fluorescence, jetness and surface structure or grain measurements. Our visual quality perception is influenced by color, gloss and the surface structure or grain. Our visual rating takes all 3 parameters into consideration and makes an overall judgement. Up till now, grain or surface structure could only be assessed visually or with highly sophisticated microscopes. This has changed with the new spectro2profiler. Like our total visual judgement the new spectro2profiler measures all 3 parameters simultaneously in a robust, portable tool. The advantages of this device for quality control at paint & plastic manufacturers and users are also discussed as well as the instruments limitations and testing conditions. The significance of instrument repeatability, inter-instrument agreement and precision is emphasized as well as the advantages of possible use of digital standards related to high precision of the instruments. Measurement results are collected and analysed in modern QC software packages. Robotic QC solutions become standard in some industries such as automotive sector, and are gaining ground in many others.

"New developments in maximising the optical efficiency of Titanium Dioxide in Matt Water Based Coating systems."

FP-Opacity Pigments™ have become an established and proven way to optimise TiO_2 use in matt paints without compromise to opacity and mechanical properties. This paper will show and explain in detail ,the methodology of FP-Opacity Pigments™ technology whereby the distance between TiO_2 particles is actively controlled, thus improving the scatter per unit TiO_2.
New advances in this technology have enabled FP-Pigments to increase the air void content of the composite pigment allowing for further optimisation of High PVC Matt paint formulations, while maintaining or improving the level of scatter.

This paper will further show how the new composite performs compared to existing opacifiers providing cost effective opacity, quality and performance.

Photocatalytic active coatings for air purification

Dr. Christina Bauder, Dr. Ulrich Christ

Fraunhofer IPA, Stuttgart, (Germany), christina.bauder@ipa.fraunhofer.de

Air pollution is a huge problem nowadays. Exhaust emission in large cities, caused by automobile traffic, industrial and household heaters, leads to high concentrations of nitrogen oxides and volatile organic compounds (VOC). To reduce the concentration of harmful air pollution, both, the reduction of exhaust emission, and air cleaning installations are needed.

However, not only outdoor air pollution is a problem, also many interior spaces are contaminated with harmful gaseous pollutants. Especially in industrial factory buildings, a high concentration of harmful substances can be present. To fulfil the legal restriction of exposure limits, the purification of the contaminated air is necessary.

Photocatalytic active coatings are materials, which can be used for purifying the air. These coatings have the purpose to transform harmful gaseous pollutants in harmless degradation products via catalytic conversion under light radiation. As well, indoor and outdoor applications are possible with this technology. For outdoor applications, the presence of sun light and humidity is necessary to activate the catalytic process. Indoor applications require special light sources (e.g. UV-light).

In various research projects, photocatalytic active coatings were formulated and investigated at Fraunhofer IPA. The aim of the projects is to generate know-how of the function and efficiency of photocatalytic coatings. Another aspect is to improve the efficiency and stability of the coatings for long-term application. To get a better insight in the mechanisms of photocatalytic activity of coatings, commercially available coatings and newly developed coatings were examined.

Different test methods were used to investigate the photocatalytic effectivity of coatings. One of these methods is the NOx-degeneration according to ISO 22197-1. Another applied method is the investigation of the degeneration of model substances, e.g. isopropanol, in the FPL-reactor test. In this test reactor, the kinetics of photocatalytic degeneration of VOCs can be investigated. Both methods are online measurement systems and give a deep insight in the mechanism and kinetics of the catalytic degeneration of the new photocatalytic active coatings. In addition, the methylene blue degeneration according to DIN 52980 is another method to quantify the photocatalytic activity of coatings.

The stability of photocatalytic active coatings often is a problem, therefore, investigations of the structure of fresh and long-term tested coatings give information about potential degeneration processes in the coating. These investigations are performed with microscopy and SEM to get better insights in the morphology and structure of the coatings. This helps to understand the photocatalytic mechanisms and to design new coatings with a higher efficiency and stability.

Aqueous silicone-acrylic/styrene and acrylic/styrene-silicone dispersions with hybrid particle structure for coating applications

J. Trzaskowska[1], J. Kozakiewicz[1], W. Domanowski[1], A. Kieplin[2], I. Ofat-Kawalec[1], J. Przybylski[1], H. Kuczyńska[3], M. Woźniak[2], D. Witwicki[2], K. Sylwestrzak[1]

1 ŁUKASIEWICZ Research Network - Industrial Chemistry Institute (ICI), Department of Polymer Technology and Processing, Warsaw, Poland
2 D&R Dispersions and Resins Sp. z o.o., Włocławek, Poland
3 ŁUKASIEWICZ Research Network - Institute for Engineering of Polymer Materials and Dyes, Paints and Plastic Department, Gliwice, Poland

The aqueous silicone-acrylic/styrene and acrylic/styrene-silicone dispersions with hybrid particle structure were developed as binders for architectural water repellent paints with enhanced hydrophobicity and water vapor permeability [1-3].
Synthesis of dispersions was accomplished by emulsion polymerization of silicone monomers in starting aqueous acrylic/styrene dispersion, or by emulsion polymerization of acrylic/styrene monomers in starting aqueous silicone resin dispersion.
The hybrid structure of the dispersion particles was identified by transmission electron microscopy (TEM) and confirmed by DSC and MDSC measurements.
The properties of coatings, films and selected paints obtained from hybrid dispersions, e.g. water and impact resistance, water vapour permeability, contact angle and surface free energy were investigated and were found to be dependent on composition of silicone-acrylic hybrid polymer.

[1] J. Kozakiewicz, I. Ofat, J. Trzaskowska, Silicone-containing aqueous polymer dispersions with hybrid particle structure, *Adv. Colloid Interface Sci.* 223, 1–39, (2015)
[2] J. Przybylski, J. Kozakiewicz, J. Trzaskowska, W. Domanowski, A. Kieplin, I. Ofat-Kawalec, M. Woźniak, D. Witwicki, K. Sylwestrzak, Studies on synthesis and characterization of aqueous hybrid silicone-acrylic and acrylic-silicone dispersions and coatings. Part I, *Coatings* 9 (25), 1-20, (2019)
[3] J. Kozakiewicz, J. Trzaskowska, W. Domanowski, A. Kieplin, I. Ofat-Kawalec, J. Przybylski, M. Woźniak, D. Witwicki, K. Sylwestrzak, Studies on synthesis and characterization of aqueous hybrid silicone-acrylic and acrylic-silicone dispersions and coatings. Part II, *Progress in Organic Coatings* 138 (2020), 105297

A Holistic Concept for Coating Monitoring and Maintenance on Offshore Wind Power Constructions

Sascha Buchbach, IFAM
Dr. Andreas Momber, Mühlhan AG
Dr. Volkmar Stenzel, IFAM

Fraunhofer IFAM, Bremen, Germany, e-mail: Sascha.buchbach@ifam.fraunhofer.de

The presentation is concerned with a nationally funded R&D project "RepaKorr". This project involves the development and implementation of an integral concept for the condition monitoring and maintenance of coatings in the atmospheric, splash water and tidal zone of offshore wind power constructions. The paper starts with a description of the general concept and the corresponding working steps, namely inspection with aerial devices; data acquisition and processing; condition assessment and damage classification; resulting maintenance procedure; surface preparation; application of a repair coating system; testing and certification of the repair system; quality control and documentation.

DoE (Design of Experiments) is used in order to design the tests and to evaluate the effects of the influencing parameters statistically. The ISO 20340 offshore testing scenario is utilized for the acceretaed ageing of the repair coatings

Particular attention will be spent on the first experience with the application of the aerial inspection device, the damage reporting system, the development of the repair coating system and the modified OWEA testing scenario for qualification.

Inspection

Surface Preparation

Coating Application

Effect of coatings with respect to powder wall friction

Christof Lanzerstorfer

University of Applied Sciences Upper Austria, Wels Austria,
c.lanzerstorfer@fh-wels.at

In the engineering of equipment for processing, conveying, dosing and storage of bulk solids, wall friction between the granular material and the wall material is an important factor. In the selection of the wall material, a small friction angle is often the target.
Wall friction can be described by both the wall friction angle φ_W and the friction coefficient μ. They are calculated from the wall normal stress σ_W and the corresponding wall shear stress τ_W [1]. These parameters can be determined, for example, with an annular shear tester using a wall shear cell into which a sample of the respective wall material is fixed.
Wall friction depends on the properties of both the wall material and the bulk material. While the properties of the bulk material usually cannot be altered because of process requirements, wall friction can be considered in the selection of the wall material. Important parameters of the wall material with respect to wall friction are the hardness, the surface roughness and the chemical composition [1-4]. Coatings are often on wall materials like structural steel for corrosion protection. Since little data is available on the influence of coatings on powder wall friction, this is not always considered during the design phase of the equipment. As a result, problems with the material flow can occur during operation.
In this study, wall friction angles were measured for a variety of powders and granular materials under various conditions. Structural steel was used as the wall material. The measured wall friction angles for uncoated steel were compared with the wall friction angles measured for steel coated with two different organic coatings and one inorganic coating. For dry powders, wall friction usually was higher for coated steel compared to the uncoated material. However, for powders with an increased water content the wall friction angle was the same or even lower for the wall material with an organic coating.

References

[1] Schulze, D. Powders and Bulk Solids - Behavior. Characterization. Storage and Flow. Berlin: Springer, 2007.
[2] Lanzerstorfer, C. Influence of the surface roughness of grit-blasted steel on the wall friction angle of limestone powder. Proceedings 14. Minisymposium Verfahrenstechnik, 4.-5. April 2018, Linz, Oral 6.3, 1, 2018.
[3] Heinrici, H. Wie lassen sich Siloschäden vermeiden? VDI Bericht 1918, 117–130, 2006.
[4] Heinrici, H. Jacob, T. Die schüttgutmechanische Betrachtung der Siloauslegung und ihre Anwendung auf ein Silo unter Tage. Schüttgut 11, 430–435, 2005.

Ablative polymer composite for thermal protection systems of steel combustion chambers in solid rocket motors

M. Elkady, P. Loboda, V. Shemet

National Technical University of Ukraine "Igor Sikorsky Kyiv Polytechnic Institute", Kyiv, Ukraine,
melkady35@gmail.com

Ablative carbon fiber reinforced polymer (CFRP) for thermal protection system (TPS) and the method of fixing it on structure elements of steel combustion chambers in solid rocket motors with short-time operation were developed.

CFRP of thermal protection system consists of silicon-modified phenol-formaldehyde resin (20 wt.-%), reinforced with carbon fibers (10 – 600 µm) (50 wt.-%) and filled with hollow glass microspheres from sillimanite (Ø 30 – 150 µm) with melting temperature ~ 1800 °C (30 wt.-%). The TPS with high ablative resistance and high controlled density (1,597 g/cm^3) was made of bulk moulding compound using hot compression technology with 200 °C under pressure 4 MPa and 10 minutes holding time. Thermal tests on TPS samples were carried out on cylindrical specimens (Ø 17 x 3 mm) by keeping them in a flame flow of 1100 °C for 20 seconds.

Thermal stability and ablation resistance experiments showed that the TPS samples improved with adding hollow glass microspheres (30 wt.-%) have 20,7 % weight loss under thermal test conditions. It was found that TPS "effective" thermal conductivity coefficient was 0,078 W / m×K.

The thermal protection system (Ø 17 x 3 mm) was fixed on steel substrate (Ø 17 x 3 mm) using developed adhesive based on silicon-modified phenol-formaldehyde resin reinforced with carbon nanotubes (5 vol.-%). The adhesive layer was applied on thermal protection system at the moment when the latter was being formed. Thickness of adhesive layer between the TPS and the steel substrate was ~ 20 µm.

The 1100 °C heat flow supply on samples from the thermal protection system side showed that the steel substrate surface temperature rose from 23 °C to 59 °C for the entire processing period (20 seconds).

SEM analysis showed that after thermal tests the TPS was a net formed from carbon fibers and cracked hollow glass microspheres linked together with thermal destruction mix of char residue and silicon oxide. TPS thickness cross-section showed partial spallation and peeling from steel substrate.

It was shown on flat models that cracking in thermal protection system cross section was caused by the thermal expansion values difference between TPS and steel substrate. Cracks did not extend across the whole thermal protection system width and were shaped at an angel 45° to the steel substrate surface. Cracks stopping on the hollow glass microspheres surface proves that the developed ablative TPS has high thermal resistance value and can be implemented for steel structure protection from thermal degradation in a rapid (20 seconds) hyperthermal environment (1100 °C).

On the development of new Fe-based driers for alkyd curing

Johan Bootsma, Zhi Qhi Hu, dr. Jitte Flapper, dr. Wesley R. Browne, dr. Bas de Bruin

Van 't Hoff Institute for Molecular Sciences, Amsterdam, (the Netherlands),
e-mail: j.bootsma@uva.nl

Paints and coatings play an important role in the protection and decoration of structural objects, increasing the lifetime of metallic and wooden structures in particular. One class of binders are alkyds, which are polyesters substituted with unsaturated fatty acids. Catalysed by driers based on cobalt, vanadium, iron and manganese,[1] during curing the unsaturated fatty acid tails are crosslinked in a radical-based process involving oxygen from the air,[2] leading to a hardened coating.

Currently the main drier used for alkyds is based on cobalt, which uses methyl ethyl ketoxime (MEKO) as an additive to prevent skinning to take place during storage. In the last years many new iron- and manganese-based driers have been developed,[3] primarily in a drive for compliance with changing legislation such as the European REACH programme (http://www.cobaltreachconsortium.org/). For iron-based driers, however, an important aspect of alkyd curing in which they fall short in comparison to cobalt- and manganese-based driers is their lack of latency.[4]

Here we show that an alternative, MEKO-free method allows for the introduction of latency in the case of iron-based driers as well. Furthermore, using these driers the coatings show an increase in hardness over time similar to cobalt-based systems, which is rarely observed for cobalt-free systems. The curing process has been studied by IR spectroscopy and confocal Raman microscopy, which show similarities as well as differences with respect to the curing process when using known driers. Lastly, the fate of the iron-based drier has been studied using UV-vis and EPR spectroscopy.

References:

[1] Soucek, M.D., Khattab, T., Wu, J. Review of autoxidation and driers. *Progress in Organic Coatings.* 73 (4), 435 – 454, (2012)

[2] Muizebelt, W.J., Nielen, M.W.F. Oxidative Crosslinking of Unsaturated Fatty Acids Studied with Mass Spectrometry. *Jounal of Mass Spectrometry.* 31 (5), 545 – 554, (1996)

[3] Simpson, N., Maaijen, K., Roelofsen, Y., Hage, R. The Evolution of Catalysis for Alkyd Coatings: Responding to Impending Cobalt Reclassification with Very Active Iron and Manganese Catalysts, Using Polydentate Nitrogen Donor Ligands. *Catalysts.* 9 (10), 825, (2019)

[4] Hage, R., de Boer, J.W., Maaijen, K. Manganese and Iron Catalysts in Alkyd Paints and Coatings. *Inorganics.* 4 (2), 11, (2016)

Anti-graffiti properties depending on the value of the free surface energy of coatings

Małgorzata Zubielewicz, Ewa Langer, Marzena Nowicka-Nowak

[1]*The Łukasiewicz Research Network – Institute for Engineering of Polymer Materials and Dyes, Toruń, Poland, m.zubielewicz@impib.pl, e.langer@impib.pl, m.nowak@impib.pl*

An adhesion between contacting phases, i.e. between anti-graffiti coating and graffiti, mainly depends on the size of the adjacent surfaces and the intermolecular interaction at the interface. The surface size of the contacting phases depends on the wetting ability of the surface with a paint, in this case the paint used for graffiti, and condition and structure of the wetted surface, in this case anti-graffiti coating. One of measures of the surface condition of polymer coatings, mainly used to study phenomena related to adhesion, is the value of free surface energy γ_S. According to the commonly accepted criterion, proper wetting of polymer surfaces occurs when their free surface energy is at least 10 mJ / m^2 higher than the surface tension of the coating material. The surface structure of a given material is also an important factor influencing the interfacial interactions. The more developed surface the better paint adhesion, the less developed surface the weaker graffiti adheres to the protected surface.

The aim of the study was to determine at which limit values of free surface energy of coatings with anti-graffiti properties, inscriptions and drawings are easily removed from the coating without damaging it. It was found that the value of free surface energy cannot be a criterion for classifying paints or varnishes to the group of coating materials with anti-graffiti properties, because the ease of graffiti removal is not directly dependent on it. In a similar way, graffiti is removed from coatings with high and low free surface energy. It seems that the removal of graffiti is more influenced by the structure of the coating – its development, microporosity, etc.

High Throughput meets Machine Learning - "Digital Coatings" for sustainable products and applications

L.Wagner[1], C.Stoffelen-Janssen[1], C.Schmitz[1], J.Göttert[1], K.Cremanns[2],

[1]Institute for Surface Technology (HIT), Hochschule Niederrhein, Krefeld, Germany
[2]PI Probaligence GmbH, Augsburg, Germany

lasse.wagner@hs-niederrhein.de

Sustainability and circular economy are two major trends in the paint and coating industry. Automation and digitalization are two often mentioned tools to quickly respond to these challenges as part of an innovative transformation towards sustainable industrial solutions.

Besides the ILOC Institute the University of Applied Sciences Niederrhein has complemented its R&D methods for developing paints and coatings with the HIT Institute to improve support and partnership for academic and industrial partners. A state-of-the-art high-throughput system automatically formulates, applies, and characterizes novel coating materials. This automation system is continuously supplemented by customized, (semi-)automated modules to replace manual tests and improve the quality of measurement and data.

To accelerate parameter screening and to minimize the number of experiments, an intelligent Design of Experiment (iDoE) is employed to simultaneously optimize the formulation recipe as well as processing parameters. During this process a machine learning model of the recipe is developed, which is the digital twin for this particular product [1,2]. This methodology is called iHIT Solution Engine.

In our poster presentation we are show-casing how the methodology works in practice. Based on our introduction to the hardware, we will discuss challenges, issues and requirements for the chemicals used. Especially dosing of liquids and powders, mixing and application such as drawdown and spray coating is explained in detail. Operating the equipment and interfacing it with the iDoE software completes the description. To illustrate how the iHIT SE works in practice, we will show the modelling of a paint formulation with regard to optical properties such as hiding power, uniformity, color, haze, gloss. We will also discuss how the digital model evolves, which information it provides and how it is used for product optimization.

In our discussion, we address the obstacles by transferring laboratory processes to fully automated ones, time- and cost-savings, as well as opportunities for iDoE solutions. With our iHIT SE approach, we want to educate and enable our partners to integrate the digital modeling approach into their everyday R&D work for developing sustainable products and being resource efficient on demand.

References:

[1] Cremanns, Kevin & Schmitz, Christian & Wagner, Lasse. (2020). A Computer learns colors. European Coatings Journal. 07-08/2020. 34-39.

[2] Schmitz, Cremanns, Bissadi (2020). Application of machine learning algorithms for use in material chemistry in Computational and Data-Driven Chemistry Using Artificial Intelligence. doi.org/10.1016/B978-0-12-822249-2.00001-3

Versatile dispersing agents for waterborne coatings

G. Gil Alvaradejo, C. Herstad, J. Sutcliffe, M. Nygren.

Dynea AS, Lillestrøm, Norway. e-mail: gabrielagil.alvaradejo@dynea.com

Pigment dispersing agents are essential components in coatings formulations, as their main function is to overcome the attraction forces that hold pigment particles together.[1] As a result, dispersing agents greatly influence the color strength, gloss, storage stability, and opacity of paints and coatings.[2]

Our Dynoadd series of pigment dispersing agents is based on novel non-ionic, structured polymers designed for use in waterborne pigment dispersions for applications including automotive, industrial, and wood coatings, as well as printing inks.

The series of dispersing agents provides excellent pigment affinity, rapid initial wetting and deflocculation, and low viscosity, when tested by dispersing a variety of pigments (carbon black, organic, inorganic) to produce resin-free pigment pastes.

Paint samples made with the addition of 10% pigment pastes were used to assess coating properties such as color, gloss, and jetness. Compared to other commercially available dispersing agents, the coatings containing Dynoadd based pigment pastes produced films having excellent color and gloss in a wide variety of waterborne industrial coatings and printing inks, with equivalent or better overall performance.

The effective steric stabilization obtained with the series of Dynoadd dispersing additives results in stable dispersions with low particle size, displaying excellent performance across a wide variety of pigments.

References:

[1] Farrokhpay, S. A review of polymeric dispersant stabilisation of titania pigment. *Advances in Colloid and Interface Science* 151, 24–32, (2009).

[2] Agbo, C., Jakpa, W., Sarkodie, B., Boakye, A., Fu, S. A Review on the Mechanism of Pigment Dispersion. *Journal of Dispersion Science and Technology*. 39 (6), 874-889, (2018).

Andrew White, Paul Dietz – FP Pigments OY

Minimizing TiO_2 usage in Coatings systems by accurately measuring Wet Hide : Optimization whilst maintaining Wet Hide Performance

A key property of FP Pigments Opacity pigment, is the ability to maintain "wet hide" in coatings at higher TiO_2 replacement levels. In this paper we will discuss the method developed by FP Pigments to more accurately measure "wet hide" and its use in determining the impact of TiO_2 substitution with various functional extenders and alternative pigments.

The use of this method, in combination with several other important coatings properties allows for a more efficient optimization of the coating being developed and hence the minimization of TiO_2 usage without compromising paint quality.

Novel Efficient Photoinitiator Systems for Curing of Water-Based Dispersions in Combination with High-Power Near-Infrared LEDs

L. Appelhoff, B. Strehmel

University of Applied Sciences Niederrhein, Krefeld, Germany,
e-mail: lukas.appelhoff@hsnr.de

Demands of the society, industry and science have addressed new aspects in coating sciences with focus to enable more aspects of sustainability. This has become one of the greatest motivations for the development of new types of coating systems. One approach, which fits into this motivation, represents the combination of solvent-free UV-curable coatings and waterborne coatings to unify the advantages of both technologies. As a result, the use of volatile, undesirable-smelling, and low-molecular weight monomers with recognizable vapor pressure could be avoided while the coating exhibits the feature to be dilutable with water. In addition, the poor performance of non-crosslinked waterborne coatings (e. g. mechanical and chemical resistance) is improved by a curing step.

Recent investigations showed that heptamethine cyanines can generate enough heat in combination with a high-power NIR-LED. This possesses potential to replace oven systems.[1] As a consequence, the heat generated can also improve the drying speed of water-borne coatings. Moreover, heptamethine cyanines used as sensitizer can initiate free radical polymerization in combination with iodonium salts upon irradiation with near infrared radiation.[2] Here, the anion receives special attention to make the system compatible with conditions employed for the aqueous dispersion. Advantages of NIR radiation are i)less danger to human skin than UV radiation ii) deeper penetration depth due to less scattering and iii)the possibility to embed UV filter material in coatings, which would make their use impossible in UV-curing systems.

Here, we describe for the first time the successful incorporation of a NIR photoinitiator system into a water borne polyurethane dispersion. First, an already proven NIR photoinitiator system for 100% coatings worked in the water-borne coatings using a high boiling cosolvent. Different counter anions for the Bis[4-(tert-butyl)phenyl]iodonium cation were compared in combination with different heptamethine cyanines. The curing is induced with high-power NIR-LEDs emitting either at 820 nm or 860 nm with an intensity of 200 mW/cm^2. Nevertheless, the reactivity could not compete with a UV reference system comprising Isopropylthioxanthone (ITX) while the exposure intensity at 395 nm was similar; that is 200 mW/cm^2. Switching to a water soluble cyanine and a water soluble Bis[4-(tert-butyl)phenyl]iodonium salt facilitated the quantitative removal of the co-solvent. Additionally, photo-DSC and FT-IR spectroscopy facilitated to study how concentration of the iodonium salt and the cyanine influence reactivity and conversion.

References:

[1] Schmitz, C., Strehmel, B. NIR LEDs and NIR lasers as feasible alternatives to replace oven processes for treatment of thermal-responsive coatings. *J. Coat. Technol. Res.*. 16 (6), 1527 – 1541, (2019)

[2] Strehmel, B., Schmitz, C., Kütahya, C., Pang, Y., Drewitz, A., Mustroph, H., Photophysics and photochemistry of NIR absorbers derived from cyanines: key to new technologies based on chemistry 4.0. *Beilstein J. Org. Chem.* 16, 415 – 444, (2020)

In situ Raman spectroscopy and headspace analysis to identify the decomposition pathway of alkyl hydroperoxides

H. den Besten[1], L.E. Eijsink[1], J. Flapper[2], K.J. van den Berg[3], W.R. Browne[1]

1) Stratingh Institute for Chemistry, Faculty of Science and Engineering, University of Groningen, Nijenborgh 4, 9747AG, Groningen, The Netherlands, h.den.besten@rug.nl

2) Akzo Nobel Decorative Coatings B.V., and 3) Akzo Nobel Car Refinishes N.V., Rijksstraatweg 31, 2171 AJ, Sassenheim, The Netherlands

Alkyl (hydro)peroxides are widely used as initiator in radical polymerization reactions. The decomposition can be triggered by heating (thermal decomposition) or by the addition of a metal catalyst. Metal catalysts are of particular interest due to the possibility to initiate reactions at ambient temperature. The reaction kinetics observed are highly dependent on the alkenes and the catalysts used. In this contribution the effect of organic solvents on the metal catalyzed decomposition pathway of alkyl hydroperoxides is evaluated. Several manganese and iron based catalysts were studied. The reactions were monitored and breakdown products identified *in situ* by Raman spectroscopy, as well as UV-vis spectroscopy. The alkyl hydroperoxide concentration over time was determined accurately in several solvents over time, using iodine liberation methods and the oxygen concentration in the headspace of a sealed reaction container was monitored to probe whether any non-productive decomposition of hydroperoxide is occurring. In this presentation we will describe the experimental methods used and their validation.

Self-stratifying coatings based on Schiff base epoxy resins

S. Waśkiewicz[a], E. Langer[b], G. Kamińska-Bach[b]

[a]Department of Physical Chemistry and Technology of Polymers, Faculty of Chemistry, Silesian University of Technology, 9 Strzody, 44-100 Gliwice, Poland, e-mail: sylwia.waskiewicz@polsl.pl

[b]The Łukasiewicz Research Network - Institute for Engineering of Polymer Materials and Dyes, ul. M. Skłodowskiej-Curie 55, Toruń 87-100, Poland

The aim of this scientific study was to use Schiff base epoxy resin as a component of self-stratifying compositions with different acrylic resins. Schiff base epoxy resins, including epoxy and azomethine (–CH=N–) groups, are characterized by high steel, copper and aluminum substrate corrosion protection and high thermal stability, while maintaining low processing temperatures [1-3]. The main drawback of these materials is their brittleness and lack of toughness. The literature describes various ways of modifying the resins in order to improve their elasticity. Mimura et al. [4] provided elasticity diglycidyl ether of epoxy resins based on bisphenol A (BPA) using polymerized in situ thermoplastic polymers. The thermoplastic polymer was formed during the curing process of epoxy resins by radical polymerization of N-phenylmaleimide, benzyl methacrylate, and styrene monomers. We are proposing another method for obtaining flexible coating systems based on Schiff base epoxy resins. It involves the incorporation of thermoplastic acrylic resins into the coating composition. The aim of this work was to obtain multilayer coating which protects the steel substrate against corrosion (epoxy resin) as well as to have decorative properties and weather resistance (acrylic resin).

A study was carried out of the structure and properties of coatings prepared from different epoxy resins and acrylic thermoplastic polymers. The degree of phase separation depended on the physical and chemical properties of acrylic resins, especially on surface energy and the amount of epoxy resin in the coating composition. In most cases, the acrylic resin present in the composition additionally served as a plasticizer, and despite using a low molecular weight crosslinking agent (polyoxypropylenediamine, average molecular weight 230 g/mol), elastic coatings were obtained. Schiff base epoxy resin, thanks to its high surface energy (52.5 mJ/m^2), meets the conditions needed for obtaining the right level of stratification of an epoxy/acrylic resin composition. On the basis of the studies, it was found that the behavior of Schiff base epoxy resin in a coating composition is similar to that of resin based on BPA with a similar epoxy number, but higher molecular weight.

References:

[1] Atta, A.M., Shaker, N.O., Maysour, N.E. Influence of the Molecular Structure on the Chemical Resistivity and Thermal Stability of Cured Schiff Base Epoxy Resins. *Prog. Org. Coat.* 56, 100–110, (2006)

[2] Chantrasiri, N., Sutivisedsak, N., Pouyuan, C., Thermally Stable Metal-Containing Epoxy Polymers from an Epoxy Resin-Schiff Base Metal Complex-Maleic AnhydrideSystem. *Eur. Polym. J.* 37, 2031–2038, (2001)

[3] Hashi Omar, I., Zucchi, F., Trabanelli, G., Schiff Bases as Corrosion Inhibitors of Copper and Its Alloys in Acid Media. *Surf. Coat. Technol.* 29, 141–151, (1986)

[4] Mimura, K., Ito, H., Fujioka, H. Toughening of Epoxy Resin Modified with In Situ Polymerized Thermoplastic Polymers. *Polymer.* 42, 9223–9233, (2001)

Optimizing coil coating performance using additives from bio-based feedstock

Camila H. dos Santos, Gabriela Gil Alvaradejo, Charlotte Herstad

Dynea AS, Lillestrøm, Norway, camila.dossantos@dynea.com

Nowadays the necessity to replace fossil-based materials with biobased materials has increased. Consequently, the market, regulatory organizations, and institutions, demand more sustainable and environmentally friendly alternatives. This demand is present in the polymeric additive and coatings industry as well [1,2].

Therefore, this project was developed to design the synthesis parameters to produce a polymeric additive for coil coatings towards the incorporation of biobased raw materials in the synthesis.

In addition to the polymer synthesis and characterization steps, application performance of the final product was also evaluated where two market reference additives, products 1 and 2, were used as benchmarks. The synthesized polymer was incorporated into coil coating formulations to evaluate foam formation and the appearance of the films made with the paint formulation. The new bio-based additive displayed excellent performance, equivalent to, or better than the market reference products.

References:

[1] Babu, R. P., O'Connor, K. & Seeram, R. (2013). Current progress on bio-based polymers and their future trends. Progress in Biomaterials, 2 (1): 8. doi: 10.1186/2194-0517-2-8.

[2] Siyab, N., Tenbusch, S., Willis, S., Lowe, C. & Maxted, J. (2016). Going Green: making reality match ambition for sustainable coil coatings. Journal of Coatings Technology and Research, 13 (4): 629-643. doi: 10.1007/s11998-015-9766-0.

Nanomaterials in fire protection. Application and requirements.

A.Rabajczyk, M.Zielecka, T.Popielarczyk, T.Sowa
Scientific and Research Center for Fire Protection - National Research Institute, Nadwiślańska 213, 05-420 Józefów, Poland

Nanotechnology is used, to an increasing extent, in practically every aspect of the economy and society. One area where nanotechnology is constantly advancing is fire protection. Nanostructures are found in elements used in direct protection, such as in protective clothing, filters, helmets. Increasingly higher challenges in the field of safety and security are also determined by the solutions used in the textile industry, which uses nano-compounds in the production of materials used for personal protection. Nano-compounds allow the creation of materials protecting against high and low temperatures to a greater extent than materials without nano-compounds, increasing the safety and comfort of firefighters' work. Nanostructures are also used to increase protection in chemical emergency situations. It should be noted, however, that depending on the application, different nanostructures will be used in materials resistant to high temperatures, different if it is necessary to prepare materials resistant to chemical substances. In the production of nanofillers improving the properties of materials used in personal protection, the following are used, among others: montmorillonite, hectorite, bentonite and saponite, metal oxides.

Solutions in the field of nanotechnology are also used in elements reducing the fire risk and increasing the fire safety of users, such as building materials and structures, paints, coatings or sensors. Thermal resistance is also one of the important criteria in the field of building materials. It plays a particularly important role role, as high thermal resistance ensures an appropriate level of mechanical strength for a long time. One of the nanocompounds used in the materials are carbon nanotubes, which increase thermal resistance due to better distribution of heat in the mass [1-3].

However, new solutions may also pose a threat to the safety of people and the environment. As a result of operation or combustion and degradation processes, the emission of nano-substances with toxic properties may occur. Therefore, knowledge in this field is necessary, as it allows for the appropriate targeting and use of nanotechnology.

Keywords: nanomaterials, fire protection, direct and indirect protection, threats

Subsidy for the maintenance of research potential granted for 2021 by the Minister of Education and Science. Other scientific activities

References:

[1] Łukaszczuk, P. The Application of Nanotechnology in Fire Protection, *BiTP*, 42(2), 95-102, 2016. DOI: 10.12845/bitp.42.2.2016.9

[2] Moniruzzaman, M., Winey, K.I. Polymer Nanocomposites Containing Carbon Nanotubes, *Macromolecules*, 39, 5194-5205, 2006. https://doi.org/10.1021/ma060733p

[3] Zielecka, M., Rabajczyk, A., Pastuszka, Ł., Jurecki, L. Flame Resistant Silicone-Containing Coating Materials. *Coatings* 10, 479, 2020. doi:10.3390/coatings10050479

Detection of bacterial contamination of process liquids in painting processes and avoidance by means of environmentally friendly, biocide-free techniques

Markus Cudazzo, Melanie Quast, Inga Landwehr

Fraunhofer IPA, Stuttgart, (Germany), Markus.Cudazzo@IPA.Fraunhofer.de

In recent years, many health- and environmentally critical process fluids required in painting processes wer replaced by more environmentally friendly materials. Likewise, pollutants such as heavy metals or organic solvents were eliminated in many coating materials and processes. On the other hand, this positive trend leads to the fact that in many areas of the painting line bacteria feel much more comfortable, if toxic substances are no longer present. The growth of bacteria must be prevented, because it usually disturbs the coating process. Regarding especially electro-dip coating processes, but also pretreatment processes, it is important to detect and avoid bacteria at an early stage. Normally, the measurement of the bacteria contamination by means of DipSlides is carried out in the last immersion rinse, which provides only insufficient information about the real situation in the electro-dip bath. At Fraunhofer IPA and Fraunhofer IPK, in a joint collaboration with the company Advatec and with the friendly support of various automotive manufacturers, an automated measuring system was developed and tested, which is suitable for an online measurement directly at the electro-dip bath and where the measurement isn't influenced too much by disturbing variables. On the one hand, this allows the detection of the contamination, and on the other hand, it detects the effectiveness of the disinfection, for example after the addition of a certain biocide.

Concerning the development of the measuring method, a good understanding of the mechanism of microbial contamination in electro dip-coatings is required: e.g., the representatives of the *Burkholderia cepacia complex*, living in the electro dip-bath, and having an adaptation mechanism that allows it to get along with even minor nutrients. In addition, this complex is able to form biofilms very well, which makes it more resistant to the use of biocides.

With the developed setup, a production-like online measurement of the degree of microbial contamination, for example in electro dip-paint, is possible within a few hours. During the project phase, the system was tested with intentionally contaminated culture media, electro-dip coat, ultra-filtrate rinsing liquid and anolyte as well as other liquids. As a continuous operation of more than 1 year has shown, the sensors used for the application-oriented tests provide a good robustness. Furthermore, the test setup allows easy cleaning of the components being exposed to paint.

In order to minimize or completely avoid the use of biocide in the future, in combination with the developed measuring method, our ambition is to create an environmentally friendly variant for disinfection, especially in electro-dip baths, but also in other areas (e.g. pretreatment), within the scope of new future research projects. By using purely physical processes, germs should to be destroyed by means of high shear forces, pressure differences and possibly cavitation forces. These forces should not influence the paint negatively. The technique of fluid dynamic water treatment, a process already used for cooling-, circulating-, service, and rinsing water, uses high centrifugal and centripetal forces in combination with high shear and frictional forces to destroy the bacteria.

References:

[1] Gühring, I.K.: *Mikrobieller Befall von Elektrotauchlack in der Automobilindustrie*. Stuttgart, dissertation, 2000

[2] Remus, L.: *Strategien gegen mikrobielle Kontaminationen von Lackieranlagen in der Automobilindustrie*. Braunschweig, Mainz: Nünnerich-Asmus, 2011

Development of New High-Performance Photoinitiating Systems and Their Applications in Polymer Coating Industry

Joanna Ortyl[1,2], *Patrycja Środa*[1], *Dominka Krok*[1], *Patryk Szymaszek*[1], *Filip Petko*[2], *Mariusz Galek*[2]
e-mail: jortyl@chemia.pk.edu.pl

[1]*Cracow University of Technology, Faculty of Chemical Engineering and Technology, Warszawska 24, 31 155 Cracow, Poland*
[2]*Photo HiTech Ltd., Bobrzyńskiego 14, 30-348 Cracow, Poland*

Production of polymer coatings curable by ultraviolet radiation (UV) imposed on various types of surface is applicable in several types of industry such as printing, paints and varnishes, adhesives and sealants, furniture, automotive and electronics. The key to achieving high process efficiency is to use an effective initiation system. The constant challenge for scientists is to find initiators with absorption characteristics compatible with emission characteristics of commercially used light sources. In addition, the initiator must not generate a colored product after the polymerization reaction, should be efficient in initiating the reaction and be cheap to obtain.

Taking this into account, we propose *meta*-terphenyl derivatives as highly effective photoinitiating systems for various photopolymerization processes according to mechanisms such as: free-radical, cationic and hybrid. Our photoinitiating systems can to be widely used in all processes where the oxygen access to the reagents cannot be limited particularly for coating of paints and protective layers as well as in rapid prototyping techniques. Other strong points of presented technology are followings:

- much higher initiation efficiency; shortening of the photopolymerization induction period and increase of monomers photocure process speed
- increase the rate of photocurable polymer layer production lines
- optical - advantageous properties, so that they can be used as photopolymerization processes photosensitizers for many monomers (vinyl, epoxy, acrylates, methacrylates and others)
- compatibility with UV LED diodes, which secure process without ozone emission and heat release
- energy savings through an increase in the production rate

In this research we investigated several new derivatives in the role of two component initiating system for the use in the coating industry.

Acknowledgements:

This research was funded by the National Centre for Research and Developments – Project Tango 2 [Grant No. TANGO2/340229/ncbr/2017, 2017]

Dirt pick-up resistance for waterborne coatings on wooden and mineral surfaces

Marzena Nowicka-Nowak, Katarzyna Suchoń

[1]The Łukasiewicz Research Network – Institute for Engineering of Polymer Materials and Dyes, Toruń, Poland,, m.nowak@impib.pl, k.suchon@impib.pl

Dirt accumulating on wooden and mineral surfaces contributes to the negative impact on the appearance of the protected details.
The aim of the study was to determine the composition of ecological waterborne varnish products for mineral and wooden substrates, which will reduce the ability to retain dirt in external and internal conditions.
The coatings of obtained products were characterized by reduced surface energy and their hydrophobic properties, which were the main criterion for obtaining a low level of adhesion of impurities on protected surfaces.
The results of the work allow expanding knowledge of the technology of manufacturing modern varnish products with improved utility and aesthetic qualities. During the project, the effectiveness of varnish formulation using specially selected varnish binders and appropriate auxiliaries was assessed, and the methodology for determining the tendency to dirty coatings in conditions was compared laboratory and field work. It was found that methods for assessing dirt retention trends, especially on coatings applied on mineral substrates, are not always effective due to the variety of factors operating during accelerated laboratory tests and long-term field tests in natural conditions. For mineral substrates, the results of the determination of coatings tendency for dirt retention show that, depending on the type of polymeric binder, different results are obtained. It was also noted that the criterion of reducing the surface energy of coatings is not an unambiguous determinant affecting the reduction of the coatings tendency to soiling.

Polyglycerine – catalysts for synthesis

Petr Vlasák[1], Aleš Růžicka[2], Roman Olejník[2], Jan Vrána[2]

1 - SYNPO a.s., Pardubice (Czech Republic), petr.vlasak@synpo.cz
2 - University of Pardubice (Czech Republic)

Effort to replace non-renewable resources is obvious in all parts of industry. Large biofuels production results to large amount of glycerol that is available. One way of glycerol utilization is its etherification to polyglycerol. It is used in cosmetic, food or pharmaceutical industry. Polyglycerol can be also used in coating industry as a raw material for synthesis of polyester and alkyd resins. While using of polyglycerols as additives does not require products with well-defined degree of polymerization, defined product is preferable for further synthesis.

Present processes use the catalysts and conditions leading to the product with broad molar mass distribution and relatively dark colour. Direct application of such materials without distillation, that is difficult, is therefore limited.

Our work is focus on synthesis of novel types of catalyst and the study of their efficiency in glycerol polymerization. The reaction products were analysed with the use of gas chromatography with mass spectrometer detector (GC-MS) and nuclear magnetic resonance (1H and 13C NMR). Developed analytical procedures allow to quantify conversion of glycerol, determine content of dimer, trimer and tetramers, but they also allow determination of the content of cyclic side products.

Organotin salts of triflouromethanesulfonic acid (triflates) and organotin triflimides (trifluoromethanesulfonimides) were prepared and tested as the acidic catalysts. Their in-situ preparation was also studied. Catalyst activity was compared with LiOH as a standard catalyst. It was observed that new catalysts are more efficient. Reaction temperature can be much lower (140-180°C) than in standard process (approx.250°C). Lower temperature has positive effect to extend of side reactions, especially to lower formation of cycles. In addition, prepared oligoesters does not contain any traces of toxic acrolein that is often present at acid catalysis.

Basic compounds were the second group of newly tested catalysts. Among them, the aryl/alkyl lithium compounds exhibit the highest efficiency. High conversion can be reached at 160 – 220°C.

References:

[1] N. Sayoud, K. D. O. Vigier, T. Cucu, B. D. Meulenaer, Z. Fan, J. Lai, J.-M. Clacens, A. Liebens, F. Jérôme: Homogeneously-acid catalyzed oligomerization of glycerol, *Green Chem.*, **2015**, 17, 4307-4314 (2015)

[2] Orita, J. Xiang, K. Sakamoto, J. Otera: Synthesis and characterization of organotin triflate clusters, *J. Organomet. Chem.* **2001** *624*, 287-293 (2001)

[3] D. Kansy, K. Bosowska, K. Czaja, A. Poliwoda: The Formation of Glycerol Oligomers with Two New Types of End Groups in the Presence of a Homogeneous Alkaline Catalyst, *Polymers,* **2019**, *11*, 144-155, (2019)

[4] M.V. Sivaiah, S. Robles-Manuel, S. Valange, J. Barrault: Recent developments in acid and base-catalyzed etherification of glycerol to polyglycerols, *Catalysis Today*, **2012,** *198*, 305-313 (2012)

Breaking up titanium dioxide particles in ball mill – experimental and CFD analysis

R. Krzosa[1], Ł. Makowski[1], W. Orciuch[1], R. Adamek[2]

[1]Warsaw University of Technology, Poland

[2]ICHEMAD-Profarb, Gliwice, Poland

email address of corresponding author: radoslaw.krzosa.dokt@pw.edu.pl

Titanium dioxide has many industrial and technical applications. Most commonly is used as a white paint base. Nevertheless, it has many interesting features that allow to apply it in other brands. Owing photocatalytic properties is used in water splitting reaction and decomposition of organic substances [1]. Titanium dioxide particles are also used as an antimicrobial coating. Owing very high refraction index TiO_2 is used in the optics industry[2].

Industrial grade titanium dioxide powder often has a structure of agglomerates of primary particles or aggregates. Compounds of agglomerates are connected via weak van der Waals interactions, which can be easily broken by hydrodynamics stresses. In aggregate primary particles are connected via chemical bonds. Therefore, those connections are more difficult to destroy. The desirable properties of a product depend on not only the composition but also on the preparation method.

Reduction of particle sizes can take place in various types of devices. For the deagglomeration of the particles, mixers equipped with high shear impeller [3,4], ball mills, pressure nozzles, and ultrasonication can be applied.

In the ball mill, rotation of the shaft inside the milling chamber generates movement of the filling balls. Collisions of the balls with each other and with chambers wall generate stresses that can break up titanium dioxide particles and reduce their size.

In this work, milling of titanium dioxide particles in a ball mill was obtained. Experimental results allowed to identify the mechanism of particle breakup and the influence of process parameters on the final product. Application of CFD analysis allows to understand process hydrodynamics and ball movement in milling chamber and find parameters that can be useful in population balance modelling.

[1] Fujishima, A., Rao, T.N., Tryk, D.A. Titanium dioxide photocatalysis. J. Photochem. Photobiol. C Photochem. Rev., 1-21, 2000.

[2] Yu, J., Zhao, X., Zhao, Q., Wang, G. Preparation and characterization of super-hydrophilic porous TiO2 coating films. Mater. Chem., 253-259, 2001.

[3] Krzosa, R.; Makowski, Ł.; Orciuch, W.; Adamek, R. Population Balance Application in TiO2 Particle Deagglomeration Process Modeling. Energies, 14, 3523, 2021

[4] Krzosa, R.; Wojtas, K.; Golec, J.; Makowski, Ł.; Orciuch, W.; Adamek, R. modelling particle deagglomeration in a batch homogenizer using full CFD and mechanistic models. Chemical and Process Engineering 42, 105-118, 2021

Synthesis of Formaldehyde Free and Water Resistant Poly Vinyl Acetate Homopolymers

Özlem Yılmaz Adıgüzel[1,2], *Barış Karabıyık*[1], *Mesut Eren*[1]

[1] *Betek Boya ve Kimya Sanayi A.Ş., Kocaeli, Turkey, yilmaz.ozlem@betek.com.tr*

[2]*Gebze Technical University, Kocaeli, Turkey*

Poly vinyl alcohol (PVOH) based poly vinyl acetate (PVAc) is one of the most crucial adhesive for different industrial areas because of their specific properties, e.g. low cost, ease of use, simplicity of application, and minimal harmful environment effects. PVAc is mainly preferred in the wood industry but some negative aspects limit the PVAc usage. PVOH which is used as a protective colloid, reduces water resistance due to the intense hydroxyl groups in their structure.[1] Water solubility of PVAc is higher than homopolymers of butyl acrylate (BA), methyl methacrylate (MMA), and ethyl acrylate(EA). There have been different approaches to improve water resistance of PVAc, for example copolymerization with less hydrophilic monomers e.g. BA, MMA, blending with compatible resins such as melamine urea-formaldehyde, adding transition metal to forming complexes with PVOH, adding internal cross-linking agent, usually N–methylolacrylamide (NMA) during polymerization, and adding inorganic salts, such as aluminium chloride.[2]

In this study, water resistant, formaldehyde- and inorganic salt- free PVAc homopolymer was synthesized with using different amount of vinyl silane functional monomer. Solid content (%), viscosity, glass transition temperature (Tg), pH, termogravimetric analysis (TGA), water absorptions, mechanical tests (TS 3891) and corrosion levels are evaluated.

References:

[1] Lijun Q., Allan J. E., Clive J. B., Philip K. C., and Robert A. F., Improvement of the water resistance of poly(vinyl acetate) emulsion wood adhesive, Pigment & Resin Technology,29(3), p.152-158, (2000)
[2] Fabio C. , Benedetto P., Innovation in poly(vinyl acetate) water resistant D3 glues used in wood industry, International Journal of Adhesion & Adhesives, 70, p.102-109, (2016)

(Poly)siloxanes as precursors of anti-fog materials

J. Karasiewicz, Z. Janicka, H. Maciejewski

Adam Mickiewicz University in Poznań, Faculty of Chemistry, ul. Uniwersytetu Poznańskiego 8, 61-614 Poznań, Poland, e-mail: joanka@amu.edu.pl

Haze can be defined as a natural phenomenon that occurs when moist air condenses and transforms into small and discrete liquid drops on a solid surface whose temperature is less than or equal to the dew point. The resulting drops form an irregular pattern that causes a blurred view due to the scattering of incident light in all directions. As a result, optical properties such as the ability to transmit lightthrough transparent materials are severely limited. Although condensation is not harmful, its occurrence annoying, because it usually arises on objects used in our everyday lives, such as, bathroom mirrors, eyeglasses, swimming goggles, binoculars, glass lenses or camera lenses, to name a few [1-2].

Superhydrophilic and superhydrophobic surfaces have aroused great interest, because they can effectively alleviate the onerous effects of fogging by simply changing the droplet morphology. The principle of the superhydrophilic surface is based on the use of materials that interact strongly with water droplets creating a continuous or amost continuous thin layer of water on the solid surface.

Effective synthesis of polysiloxanes containing mixed functional groups was carried out using consecutive hydrosilylation of two diferrent olefins. The functionalization of polysiloxanes was aimed at obtaining derivatives containing fluoroalkyl groups (having hydrophobic properties) and polyether groups (having hydrophilic properties). All derivatives obtained were isolated and characterized spectroscopically. In the next stage of research, the glass plates were modified with functionalized polysiloxanes and their anti-fogging properties were checked.

By modifying, by attaching different groups to the main polysiloxane chain, materials with a wide range of applications can be obtained. The communication will present the results of surfaces modified with polysiloxane containing mixed functional groups and the best anti-fog coating will be selected.

References:

[1] Rodríguez Durán I., Laroche G. *Advances in Colloid and Interface Science* 263, 68–94 (2019)
[2] Wen S.F., Wang Y.M., Zhang Z.M., Liu Y.L., *Progress in Organic Coatings* 137, 105320(2019)

Bifunctional organosilicon compounds and their potential application

J. Karasiewicz, Z. Janicka

Adam Mickiewicz University Poznań, Faculty of Chemistry
ul. Uniwersytetu Poznańskiego 8, 61-614 Poznań, Poland
e-mail: joanka@amu.edu.pl

Production of advanced materials contributes to driving social and economic development, which in turn demands new materials of target properties to meet specific needs. New materials of strictly defined structure and properties are designed on the basis of results of fundamental studies. Recently, the group of materials showing amphiphilic properties has become of increasing interest in the global search for materials of specific properties [1-2]. Compounds with amphiphilic properties are very dynamically developing, multifunctional, because they are a compilation of chemical, biological and physical research. There is a wide range of applications of this group of compounds, mainly in various industries as: washing and cleaning agents, cleaning agents, foaming agents or foam stabilizing agents, dispersing and emulsifying agents or pesticide components (adjuvants) [2-3]. The most important thing is that their specific structure makes them surface active. They can be used to obtain anti-fouling or self-cleaning protective coatings for example to wood, glass or diferennt building materials [4,5].
In this presentation, the syntheses functional siloxanes containing mixed functional groups at different stoichiometric ratios, as well as their selected surface properties were presented.

References:

[1] Kashapov R., Gaynanova,G., Gabdrakhmanov, D. Kuznetsov, D., Pavlov, R., Petrov, K., Zakharova, L., Sinyahin, O. : Self-Assembly of Amphiphilic Compounds as a Versatile Tool for Construction of Nanoscale Drug Carriers. *International Journal of Molecular Science*, 21, 6961-7008, (2020);

[2]Ghosh, S., Ray, A., Pramanik, N. Self-assembly of surfactants: An overview on general aspects of amphiphiles. Biophysical Chemistry, 265, 106429- 106440 (2020);

[3] Khan, M.F., Zepeda- Velazquez, L., Brook, M.A., Tunable, antibacterial activity of silicone polyether surfactants. *Colloids and Surfaces B: Biointerfaces*, 132, 216- 244 (2015);

[4] Guazzelli, E., Martinelli, E., Pelloquet, L., Briand, J-F., Margaillan, A., Bunet, R., Galli, G., Bressy, C.: Amphiphilic hydrolyzable polydimethylsiloxane-*b*-poly(ethyleneglycol methacrylate-*co*-trialkylsilyl methacrylate) block copolymers for marine coatings. II. Antifouling laboratory tests and field trials. *BIOFOULING*, 36, 378-388 (2020);

[5] Adamopoulos, F.G., Vouvoudi, E.C., Achilias, D.S., Karapanagiotis, I.: Fluorosilane Water-Repellent Coating for the Protection of Marble, Wood and Other Materials. *Heritage* , *4*, 2668-2675 (2021)

Photo-crosslinking polymers by dynamic covalent disulfide bonds

Bianka Sieredzinska[1], *Qi Zhang*[1], *Keimpe van den Berg*[2], *Jitte Flapper*[3], *Ben L. Feringa*[1]

[1]*University of Groningen, Groningen, the Netherlands,*
e-mail: b.e.sieredzinska@rug.nl
[2]*Akzo Nobel Car Refinishes B.V., Sassenheim, the Netherlands*
[3]*Akzo Nobel Decorative Coatings B.V., Sassenheim, the Netherlands*

Cross-linking polymers by light is of major importance for numerous applications ranging from coatings to 3D-laser printing. Commonly, photo-initiators are used for irreversible polymerization and cross linking. However, a major challenge in view of environmental and sustainability issues is the development of polymeric materials with structurally simple units with intrinsic cross-linking capacity and dynamic covalent bonds. Seemingly contradicting, it should support a robust cross-linking network as well as maintaining intrinsic reversibility i.e. photocross-linking and de-crosslinking under mild conditions while not compromising materials properties.

A simple and general strategy to construct photo-crosslinkable polymers by introducing sidechain 1,2-dithiolanes based on natural thioctic acid is presented[1][2][3][5]. The disulfide five-membered rings act both as light-absorbing and dynamic covalent crosslinking units, enabling efficient photo-crosslinking and reversible chemical decrosslinking of polydimethylsiloxane polymers[4][6].

References:

[1] Barltrop, J. A., Hayes, P.M, Calvin, M. *J. Am. Chem. Soc.*, 76, 4348-4367, (1954)
[2] Reed, L. J., DeBusk, B. G., Gunsalus, I. C., Schnakenberg, G. H. F. *J. Am. Chem. Soc.*, 73, 5920 (1951)
[3] Schmidt, U., Grafen, P., Goedde, H. W. *Angew. Chem. Int. Ed.*, 4, 846-856 (1965)
[4] Zhang, Q., Shi, C. Y., Qu, D. H., Long, Y. T., Feringa B. L., Tian, H. *Sci. Adv.*, 4, eaat8192, (2018)
[5] Calvin, M. *J. Chem. Phys.*, 23, 1750, (1955)
[6] Shi, C. Y., Zhang, Q., Tian, H., Qu, D. H. *SmartMat*, 1, e1012, (2020)

NIR-Absorber for Photonic Drying of Coatings with High-Power NIR-LEDs between 800-950 nm

Sascha Driesen[1], Alfred Feilen[2], Bernd Strehmel[1]

[1]Niederrhein University of Applied Science, Faculty of Chemistry and Institute for Coatings and Surface Chemistry, Adlerstr. 1, D-47798 Krefeld, Germany, [2]Easytec GmbH, Pascalstr. 6, 52076 Aachen (sascha.driesen@hs-niederrhein.de)

NIR absorbers based on functionalized polymethines possess the potential to replace old oven techniques just by excitation with a photonic source; that is a high-power NIR-LED emitting between 750-1050 nm. These new devices have received great attention to solidify varnishes in combination with cyanines.[1] This worked well with devices in the range between 800-900 nm.[1]

NIR-absorbers possess the feature to convert the energy absorbed into heat and radiation. They worked well for laser drying of offset printing inks where most of the heat released resulted in physical drying of the ink applied. Thus, it can be seen as a certain challenge to combine efficient NIR-sources with respective cyanine absorbers to develop systems that solidify different paints and coatings on a chosen substrate. Here, high-power NIR-LEDs have moved into the focus since their uses offers many advantages compared to respective NIR-lasers. This work preferentially focusses on LEDs emitting either at 860 nm and 940 nm.

There exist special relations between the structural pattern of the cyanine absorber and the competition between the release of radiation and non-radiative deactivation. For this purpose, internal conversion between the lowest vibrational mode of the excited state with a higher vibrational mode of the ground state can be seen as the main source for heat generation. Collision with matrix molecules of such a hot molecule forms the main source to generate heat in the matrix. Though, the temperature generated can reach more than several hundred Celsius degree, this technology also works well on temperature sensitive substrates because light absorption and heat generation proceeds mainly in the coating.

The NIR-absorbers used absorb between 800-950 nm. They comprise either a Benzo[c,d]- or Benzo[e]indolium pattern and exhibit high extinction coefficients (3.3×10^5 $M^{-1}cm^{-1}$). In general, benzo[c,d]-indolium heptamethines solidify coatings by generation a of heat. This is important for coating systems where drying occurs by a chemical and/or physical reaction. One example shows the thermal initiation of a chemical reaction resulting in formation of a polyurethane, where the substrate, the isocyanate, was blocked. In that case, the isocyanate can be deblocked after photonic generation of heat; that is >170°C.

The solubility and therefore the compatibility with the matrix depicts another important parameter for photonic drying. Thus, aggregation should be avoided in such systems because this would result in a loss of the overall performance. The solubility can be controlled through the end groups of the absorber, its functional groups in the *meso*-position, and the counter ion as shown by several examples. The photonic drying of paint systems offers the great advantage that the heat required for drying proceeds directly in the coating. Another advantage is that the drying process can be started On Demand without the warm-up times required in conventional oven-drying methods.

[1] C. Schmitz, Y. Pang, A. Gülz, M. Gläser, J. Horst, M. Jäger, B. Strehmel, „New High-Power LEDs Open Photochemistry for Near-Infrared-Sensitized Radical and Cationic Photopolymerization", *Angew. Chem., Int. Ed.* **2019**, *58*, 4400-4404.

Corrosion protection for outdoor barbecue grills

Sz. Horvath[*], I. Felhosi[**], L. Molnar-Nagy[*]

**FKI Festékipari Kutató Kft.. Budapest (Hungary)*
***Institute of materials and environmental chemistry Functional Interfaces Research Group (Hungary)*
Contact: horvath.szilvia@festekkutato.hu, www.fki.eu

The market demands of carbon-steel thermal equipment (e.g. heaters, ovens, BBQs) have to handle high temperatures and resist corrosion effects.

There are heat resistant coatings up to 500°C and others for corrosion protection, but challenging to combine these two criteria in a single-layered DTM coating.

Before this work, proven heat-resistant and anticorrosive properties coatings checked by electrochemical impedance spectroscopy and salt-spray methods were studied to determine the optimal application conditions on shot blasted hot- and cold-rolled steel surfaces.[1],[2]

However, the thin cold-rolled steel sheets tend to deformation by shot blasting, so the industrial practice forgo it risking adhesion problems.

We studied the effect of the surface roughness, the layer order and composition of the applied coatings and the method of heat demand on the coating properties of heating equipment intended for outdoor use.

We compared results with properties of the conventional bilayer (Zn-rich primer and topcoat) coating system.

Both salt spray and EIS was used to monitor the electrochemical activity of the paint substrate interface.

This study aims to pay attention to the importance of surface preparation and compare thermo-resistant and anticorrosive properties of single-layered and bilayered systems through these surfaces.

[1] T. Pozman et al. Corrosion protection and heat resistance in a single coat, ETCC2018 (2018)

[2] I. Felhősi et al. Corrosion protection effects of a heat-resistant paint, Virtual EUROCORR2021, (2021)

Novel coatings for wood-based panels

Daniel Schaupp, Merete Selvåg

Dynea AS, Lillestrøm, Norway, email: Daniel.schaupp@dynea.com

A new paperless, liquid overlay technology for wood-based panels has been developed and launched as Trollon™. It is a new smart alternative to traditional coatings and paper overlays. This highly efficient single-coat technology, is applied at flexible application rates and gives board producers the opportunity to coat and finish added value boards in-house, using standard equipment for the coating and crosslinking process used already for panel production. This new, highly efficient board coating concept provides surfaces with opaque, transparent, or clear appearance that are free from fiber lifting. Trollon is a two-component system, applied by roller spreader and cured in a hot or radio frequency (RF-) press. After pressing, the boards with cured surfaces require no sanding, are stackable and ready for use.

Here we present performance optimization and determination of the cure window for a new shutterboard coating. The coating samples were tested for wet-heat resistance, concrete adhesion, elasticity, and chemical resistance and compared with well-known surface treatments for wood-based panels. Press conditions of the coating were correlated against the coating performance.

References:

[1] EN438-2, High-pressure decorative laminates (HPL) – Sheets based on thermosetting resins (usually called laminates) – Part 2: Determination of properties, ICS 83.140.20, February 2016

[2] En-13442, Wood flooring and wood paneling and cladding – Determination of the resistance to chemical agents, March 2013

[3] EN-13329, European Norm for laminate Floor Coverings, June 2010

The applications and properties of micro-arc oxidized coatings

Balbina Makurat-Kasprolewicz

Gdansk University of Technology, Gdansk (Poland),
e-mail: balbina.makurat-kasprolewicz@pg.edu.pl

Micro-arc oxidation, also known as plasma electrolytic oxidation, is a novel surface coating technique that aims to improve the properties of the material. The features that can be streamlined thanks to it comprise: wear resistance, corrosion resistance, hardness. It owes its innovation to the fact that it combines the advantages of high-temperature plasma spraying and electrochemical methods.

The materials subjected to this oxidation process include mainly aluminum, magnesium, titanium and their alloys, as well as niobium and tantalum. The final effect of the treatment can be modified by changing the voltage, electrolyte composition, treatment time, current density, etc. Hence, it can be used in biomedical engineering, aerospace engineering, electronics, ship engineering, as well as automotive engineering.

Therefore, the aim is to present the latest achievements and trends in the development of micro-arc oxidized coatings. Essentially, the focus was on characterizing the parameters, the use of which results in coatings with promising properties useful in various industries.

The methodology was based on several substitutes: technique (micro-arc oxidation, plasma electrolytic oxidation), material (aluminum, magnesium, titanium and their alloys), date of publication of the paper (from the last 2 years), alteration of material properties (wear resistance, corrosion resistance).

The analysis exhibits that titanium and its alloys are most often used in biomedical applications whereas aluminum, magnesium and their alloys in such industries as: aerospace, automobile, electronics, marine and military industries. The properties of the coatings depend on the oxidation parameters used.

References:

[1] Sikdar S., Menezes P. V., Maccione R., Jacob T., Menezes P. L. Plasma Electrolytic Oxidation (PEO) Process—Processing, Properties, and Applications. Nanomaterials. 11(6), 1375, (2021)

[2] Simchen F., Sieber M., Kopp A., Lampke T. Introduction to Plasma Electrolytic Oxidation—An Overview of the Process and Applications. Coatings. 10(7), 628, (2020)

Using IR spectroscopy and EPR spectroscopy to study a method for used rubber recycling.

Mitchell Groen, Francesco Picchioni, Wesley Browne, Martin van Duin

Faculty of Science and Engineering, University of Groningen, Nijenborgh 4, 9747 AG Groningen, The Netherlands, email: m.j.groen.1@student.rug.nl

Rubber products such as car tires prove difficult to undergo recycling processes due to their complexity. Most used rubber product are either burned for energy purposes, or ground to small particles to be incorporated in, for example, polymeric resins. High-quality recycled rubber material can be produced from recycled used rubber products for which a method was developed.[1] Thermal devulcanization is used to break the structure of a crosslinked rubber, which allows reprocessing of the rubber A branching agent re-crosslinking after processing and the process can be performed as a one pot procedure using a brabender reactor.

ATR-FT-IR spectroscopy can give insight into the chemical nature of these residues. In this presentation, the analysis of coatings on inner bike tubes (comprising of butyl rubber) to be recycled is discussed. EPR spectroscopy is used to characterise radicals formed during processing and the impact of coating additives on these radicals. The data allow for understanding the impact of coatings on processibility.

(1) Picchioni, F.; Broekhuis, A. A. Method for Reclaiming Rubber, and Renewed Rubber Compositions Obtainable Therebry. 2018, p 18.

Effect of lignin on thermal properties of chitosan and starch-based coatings processed with PVA

W. Janik, A. Wojtala, A. Pietruszka, A. Krasuska, E. Sabura

Łukasiewicz Research Network - The Institute of Heavy Organic Synthesis "Blachownia", Kędzierzyn-Koźle (Poland), weronika.janik@icso.lukasiewicz.gov.pl

Interest in biodegradable polymers has increased in recent years, mainly due to concern over the disposal of conventional synthetic plastics derived from petroleum. The degradation of plastics requires a long time and most end up in landfills. In contrast, naturally derived polymers from renewable agricultural products are biodegradable and can also be an alternative to traditional coatings. Among the biopolymers used for coatings, polysaccharides have attracted more attention due to their wide availability, biocompatibility, and the fact that they can often be obtained from waste materials [1-2]. However, polysaccharide-based materials have weaknesses and it is essential to modify them. Methods such as crosslinking or the addition of nano-additives can be found in the literature [3-4]. Whereas our results show that lignin affects the thermal properties of the obtained materials.

The aim of the study was to show the effect of different types of lignin on the thermal properties of starch- and chitosan-based coatings with PVA. The test samples were prepared by thermomechanical processing and the lignin content was 5%. The thermal analysis was conducted for a composition consisting of native potato starch, high molecular weight chitosan, PVA with a degree of hydrolysis of 71-30, acetic acid and glycerol.

In the first part of this study, the analysis was carried out using a Mettler Toledo TGA 2 Thermobalance. The samples (5 mg) were heated up in an open platinum crucible in the temperature range of 30–1100˚C with the heating rate = 20˚C/min, in dynamic nitrogen or air atmosphere. It was found that the decomposition process of all composites occurs in two stages, either in an air or nitrogen atmosphere. Moreover, a single asymmetric peak is observed in each stage, indicating the homogeneity of the composites. For the lignin-modified composites, a weight loss is observed in the temperature range of about 200˚C.

The second part of the work presents the DSC measurements, which were performed using a Mettler Toledo DSC 822e Differential Scanning Calorimeter. The measurements were carried out in two or three stages for mS ~5 mg in an aluminum crucible closed with a perforated lid, under a nitrogen flow of 50 mL/min. The glass transition temperature was determined using the enthalpy calculation procedure given by Richardson [5]. It turned out that the glass transition temperature is significantly higher for lignin than for chitosan and it was observed in the temperature range of 140-180˚C. Meanwhile, analysis of the composites showed that the thermograms exhibit deviations in the range of about 30˚C indicating the success of the plasticization process.

References:

[1] Nathanael, A.J.; Oh, T.H. Biopolymer Coatings for Biomedical Applications. Polymers. 12, 3061, (2020)

[2] Spirk, S.; Nypelö, T.; Kontturi, E. Editorial: Biopolymer Thin Films and Coatings. Front. Chem. 7, (2019)

[3] Guerrero, P.; Muxika, A.; Zarandona, I.; de la Caba, K. Crosslinking of Chitosan Films Processed by Compression Molding. Carbohydr. Polym. 206, 820–826, (2019)

[4] Abral, H.; Pratama, A.B.; Handayani, D.; Mahardika, M.; Aminah, I.; Sandrawati, N.; Sugiarti, E.; Muslimin, A.N.; Sapuan, S.M.; Ilyas, R.A. Antimicrobial Edible Film Prepared from Bacterial Cellulose Nanofibers/Starch/Chitosan for a Food Packaging Alternative. Int. J. Polym. Sci. 2021, (2021)

[5] Manley, T.R. Calorimetry and Thermal Analysis of Polymers, Edited by V. B. F. Mathot. Hanser, Munich, 377, (1994)

UKRAINIAN COATINGS MARKET : SAFETY REMAINS IN PRIORITY

Taras Karavayev

State University of Trade and Economics (SUTE), Kyiv, Ukraine;
e-mail: t.karavayev@gmail.com

Paint and coating materials (PCMs) market in Ukraine has some changes in last few years. The most significant impact was the further increase in environmental requirements and the global COVID-19 pandemic, as the reflection of global trends. In the last 3 years, important EU directives have come into force in Ukraine, which increase the safety requirements for PCMs. Legislative changes are initiated by Ukrainian Paint and Coatings Association (UPCA), which show direction of Ukrainian PCMs manufacturers on global and European trends. Ecology and safety in PCMs production and use remains quite important for Ukraine.

According to the UPCA data the total volume of PCMs production for the last 5 years was in the range from 267.6 ths tons in 2017 till 386.2 ths tons in 2019. In monetary terms, it was range of 300-425 million USD.

The following trends have been observed in the structure of production by types of PCMs for the last 5 years. The production of solvent-borne (SB) PCMs (63.4–77.4 ths tons) shows a constant downward trend, which corresponds to the global trend.

Production of water-borne (WB) PCMs in 2017-2019 increased about 11% from 103.7 to 115.2 ths tons. In 2021 production increased by 8.6% to 107.3 ths tons, which is in line with the global trend supported by leading Ukrainian manufacturers. The share of WB PCMs in the overall structure of production was in the range of 31-39%, and in the structure of production of varnishes and paints - 57-63%.

The Technical Regulation on the Limitation of Emission of Volatile Organic Compounds (VOCs) in PCMs had a significant impact on the positive dynamics of growth in the production and consumption of WB PCMs in Ukraine. This document, which entered into force on May 30, 2019, is based on the European Directive 2004/42/EC. Also positive is the entry into force in Ukraine in 2016 of almost all European standards (EN) of the series "Paints and varnishes", especially for WB PCMs. Thus, the production of WB PCMs in Ukraine carried out according to the EU requirements.

The Ukrainian market formation is also influenced by the PCMs imports, the total volume of which over the past 5 years amounted to almost 210 ths tons and ranged between 38.7 and 48.0 ths tons with a slight upward trend in both main groups (SB and WB). The volume of exported PCMs from Ukraine was 5.4–6.4 times lower than imported and was in the range of 6.4 ths tons in 2017 till 8.3 ths tons in 2021. SB PCMs are also dominated in the structure of exports.

Consumption of PCMs in Ukraine ranged from 304.1 ths tons in 2017 till 421.5 ths tons in 2019. In 2020, there was a decline in consumption by almost 15% to almost 360 ths tons, mainly due to SB and other PCMs. Consumption remained at the same level in 2021, but the increase in WB PCMs was 8.2%, in relation to SB PCMs – only 0.8%, and other PCMs – decreased by 7%. These data are another confirmation that the consumption and market of PCMs of Ukraine is developing in the direction of improving the environmental friendliness and products safety to consumers that meet global trends.

The presented data confirm the constant trend of increasing environmental friendliness and safety in the production and consumption of PCMs in Ukraine, which is in line with global and European trends. The big PCMs producers, that are members of Ukrainian Paint and Coatings Association, follow and support such environmental friendly and safety trends.

The Design and Synthesis of New NIR-Active Sensitizers for Use in Photochemical Processes and Controlled Polymerizations

Nicolai Meckbach[1)], Sascha Driesen[1)], Bernd Strehmel[1)]

[1)]Niederrhein University of Applied Sciences, Department of Chemistry, Institute for Coatings and Surface Chemistry, Adlerstr. 1, D-47798 Krefeld, Germany
nicolai.meckbach@hs-niederrhein.de

In controlled living polymerization techniques, atom transfer radical polymerization (ATRP) is the most used strategy for the targeted synthesis of tailor-made polymers that have received particularly much interest for coatings as well. The requirement to use transition metal complexes, which must have two stable oxidation states with no formation of metal oxides, depicts a major disadvantage. A successful developed and applied strategy generates in-situ the respective transition metal complex by photochemical reduction with photosensitizers. A system successfully developed in the past, which reports on the NIR-sensitized photo-induced atom transfer radical polymerization[1], was transferred to a new transition metal/catalyst system.

NIR-active sensitizers were essentially developed for these photopolymerization processes. They based on heptamethines absorbing in the range between 700 and 1300 nm. Here, we successfully synthesized different sensitizers comprising either a benzo[c,d]indol-2(1H)-one or benzo[e]indol-2(1H)-one moiety. Functionalization influenced solubility in organic surroundings or aqueous environment. The absorption maxima of these organic sensitizers depend on the chain building agent for the polymethine group, which was used throughout the synthesis process. This can be also influenced by different functionalization either. Addition of metal ion also resulted in formation of a new absorption band, which can show either a bathochromic or hypsochromic shifted absorption while the absorption of the sensitizer decreased. Metal ions used for this purpose were taken from Cu(II), Fe(III), Pt(IV), Au(III) and Ni(II) bromides. The fact that metal ions can shift the overall absorption enables them to initiate polymerization at distinct wavelengths. This brings access to new NIR light sources and the motivation to synthesize such sensitizers on respective demands.

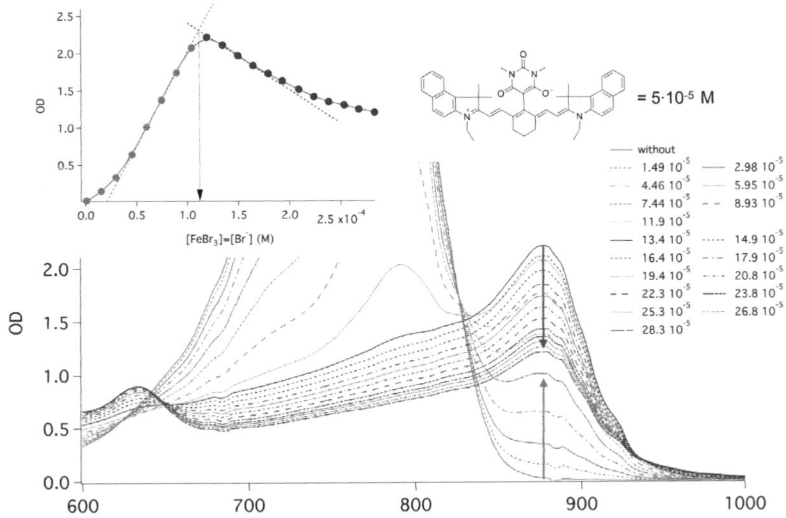

[1] C. Kütahya, N. Meckbach, V. Strehmel, B. Strehmel, „Cyanines comprising barbiturate group facilitate NIR-light assisted ATRP under anaerobic and aerobic conditions at two wavelengths using Fe(III) catalyst“, *Journal of Polymer Science* **2021**, *59*, 2023-2035.

Alkoxysilane-grafted polybutadienes as effective coating materials

A. Szymańska[1], M. Przybylak[1], M. Dutkiewicz[1,2], H. Maciejewski[1,3]*

[1]Poznań Science and Technology Park, Adam Mickiewicz University Foundation, Poznań, (Poland), e-mail: anna.szymanska@ppnt.poznan.pl
[2]Centre for Advanced Technologies, Adam Mickiewicz University, Poznań, (Poland)
[3]Faculty of Chemistry, Adam Mickiewicz University, Poznań, (Poland)

The degradation processes of some materials, e.g. natural textiles or porous building ones are often associated with the destructive action of water. Pristine cotton fabric is characterized by high polarity and hydrophilic character, therefore its usefulness in many areas may be limited by the poor water and dirt repellency, growth of microorganisms on its surface and finally weakening of cotton fibers [1,2]. In the case of concrete, penetration of water and aggressive chemical ions into micro-cracks, cracks and other defects influencing its physical and chemical degradation processes [3]. On the other hand, the presence of OH groups on the surface of cotton and concrete enables their chemical modifications towards obtaining water resistance.

The application of protective coatings precursors containing SiOR groups, which are very reactive and form chemical bonds with OH groups present on the surface of modified materials or react with each other to form siloxanes, has been suggested as a solution for the protection of aforementioned materials.
In this communication, our recent research on the preparation of functional polybutadienes containing organosilicon side groups and their use as effective waterproofing coating materials will be presented.

Acknowledgments:

This work was supported by the National Science Centre Grant OPUS 2018/29/B/ST8/00913.

References:

[1] Shateri-Khalilabad, M., Yazdanshenas, M. E. One-pot sonochemical synthesis of superhydrophobic organic–inorganic hybrid coatings on cotton cellulose. *Cellulose*. 20 (6), 3039 – 3051, (2013)

[2] Przybylak, M., Maciejewski, H., Dutkiewicz, A., Dąbek, I., Nowicki, M. Fabrication of superhydrophobic cotton fabrics by a simple chemical modification. *Cellulose*. 23 (3), 2185 – 2197, (2016)

[3] S. Weisheit, S. H. Unterberger, T. Bader, R. Lackner, Assessment of test methods for characterizing the hydrophobic nature of surface-treated High Performance Concrete. *Construction and Building Materials.* 110, 145 – 153, (2016)

Optical component fabrication by UV nanoimprint through resists based on ionic liquids

Andrea Szpecht[1], Adrian Zajac[1], Katarzyna Komorowska[2,3], Marcin Smiglak[1]

[1]Poznan Science and Technology Park, Poznan, Poland
[2] Lukasiewicz Research Network - Institute of Microelectronics and Photonics, Warsaw, Poland
[3] Wroclaw University of Science and Technology, Wroclaw, Poland
Email: andrea.szpecht@ppnt.poznan.pl

Applying lithography techniques, a resist can be patterned to create a (sub)micrometer-scale temporary mask that shields certain portions of the underlying substrate during subsequent processing stages. [1] Nanoimprint litography (NIL) is one of the more popular techniques that can lead to obtaining the contours of the desired stamp or template. [2] Polymers have been known to be the most used materials as resists. Ionic liquids possess characteristic and special properties that distinguishes them from organic solvents and other chemicals. They have almost negligible vapor pressure and high thermal stability. [3]

Results presented in this work focus on obtaining new group of ionic liquids bearing one or more 4-vinylbenzyl groups in the cation structure. Such chemicals are known to undergo radical polymerization through UV irradiation. They are used in liquid form, as novel materials for nanoimprint lithography, without any additional solvents normally used with standard resist materials. This makes ionic liquids more advantageous and makes this process safe and environmentally friendly. Structures were obtained with both hard and soft mold imprints. Few different optical components are produced, including Fresnel lens.

This work was supported by the National Science Centre (Poland), project OPUS no UMO- 2019/35/B/ST8/03736.

References:

[1] Hauser H., Tucher N., Tokai K., Schneider P., Wellens C., Volk A. V., Seitz S., Benick J., Barke S., Dimroth F., Müller C., Glinsner T., Bläsi B., *J. of Micro/Nanolithography, MEMS, and MOEMS*, 14(3), 031210, (2015)

[2] Glinsner T., Lindner P., Mühlberger M., et al. *Fabrication of 3D-photonic crystals via UV-nanoimprint lithography, Journal of Vacuum Science & Technology B: Microelectronics and Nanometer Structures Processing, Measurement, and Phenomena*, 2337- 2340, (2007)

[3] Zajac A., Szpecht A., Zielinski D., Rola K., Hoppe J., Komorowska K., Smiglak M., *Synthesis and characterization of potentially polymerizable amine-derived ionic liquids bearing 4-vinylbenzyl group, Journal of Molecular Liquids*, 283, 427-439, (2019)

Development of innovative polymer dispersions with monomers from renewable sources as modifying additives for specific applications in the construction industry

Martin Kolář, Jana Machotová

Institute of Chemistry and Technology of Macromolecular Materials, Faculty of Chemical Technology, University of Pardubice, Studentská 573, CZ-532 10 Pardubice, Czech Republic, email: st51413@student.upce.cz

Polymer latexes have been used in cementitious materials to improve properties such as workability, adhesion to substrates, anti-bleeding, and the properties of hardened materials, including flexural strength, adhesive strength, ductility, cracking resistance, and durability[1]. In recent years, a growing concern for the preservation of the environment is evoking new research, focused on alternative methods for sustainable economic development. Using monomers from vegetable oils is not only connected to environmental problems, but also to the properties of the final materials[2]. Acrylated methyl oleate (AMO) was synthesized from acrylic acid and epoxidized methyl oleate obtained from rapeseed oil. Several ways of approaching the desired product were investigated. The iodine value (representing the number of double bonds) of the products was determined using the Hanus method and the approach, which yielded the product with the highest iodine value was used for subsequent research. The semi-continuous emulsion polymerization process was used to prepare three latexes based on copolymerized AMO and commercial acrylic monomers. Various ways of approaches were tried; namely a single-step polymerization and core–shell polymerization with various monomer ratios dosed in the first and second step (1:1, 1:3). A synthesis of AMO was successfully optimized for higher amounts of starting compounds, which was controlled by higher iodine values of the products. AMO was then used with other commercial acrylic monomers to prepare a polymeric dispersions. Composites consisting of latexes and Portland cement mortar were produced and their mechanical properties were tested.

Keywords: emulsion polymerization, water-based acrylic latexes, acrylated methyl oleate, Portland cement.

References:

[1] JENNI, A., ZURBRIGGEN, R., HOLZER, L., HERWEGH, M. Changes in microstructures and physical properties of polymer-modified mortars during wet storage. *Cement and Concrete Research*. 36(1), 79-90, (2006)

[2] NEVES, J. S., VALADARES, L. F., MACHADO, F. Tailoring Acrylated Soybean Oil-Containing Terpolymers through Emulsion Polymerization. *Colloids Interfaces*. 2(4), 1-3, (2018)

How can organic coatings stability be traced with modern analytical technique: FTIR spectroscopy and electrochemical Scanning Kelvin Probe

J. Jurek-Suliga, S. Golba, J. Kubisztal

University of Silesia, Faculty of Science and Technology, Institute of Materials Engineering, Chorzow (Poland), 41-500 Chorzów, 75 Pułku Piechoty 1a, e-mail: justyna.jurek-suliga@us.edu.pl

Multilayer organic coatings deposited on a steel substrate are widely used in the automotive industry. These coatings, independently on decorative aspects, create a physical barrier between an aggressive environment and the surface of protected material and in this way increase corrosion resistance of the coating/steel system. However, one should bear in mind that the ability of a solution to penetrate an organic coating lead to possibility of forming discontinuities in the surface. With the increase in the extent of crack formation one can observe the breaking of the coating what inevitably provokes an increase of corrosion of the substrate.

The aim of the current study was to determine quantitatively the stability of the organic coatings deposited on steel substrate immersed in salt solution namely 5 wt.% aqueous NaCl solution with adjusted pH value (either to acidic (3.5) or neutral (7.0) value). The progress of the solution penetration ability was traced within 14 days.

The chemical composition of the as-received, commercially available, coatings was determined using FTIR spectroscopy. It revealed that all samples contained epoxy and phthalic resin with the addition of fillers, which increase the protective properties of the coating. FTIR spectra obtained for samples immersed in the aggressive environments do not reveal new absorption bands regardless of the pH value used. Moreover, it was shown that the position and transmittance of the characteristic peaks obtained for the organic coatings before and after immersion in the aggressive environment are similar.

The quantitative analysis of the contact potential difference (CPD) maps obtained by means of Scanning Kelvin Probe method indicates a significant decrease in the mean value of CPD by c.a. 190 mVKP. For the immersed coatings five times higher CPD fluctuation around the average value was also registered.

In general, the obtained results show that:

i) the corrosive environment causes the organic coating degradation,

ii) superficial cavities and blisters affect surface electrical properties, which influence on protection efficency

iii) the studied coatings preserve the structural stability in the measured time range.

Keywords: organic coatings stability, corrosion resistance, solution penetration ability, FTIR spectroscopy, Scanning Kelvin Probe

Digital Database Structures and Workflows for Intelligent Automation

C. Schmitz[1], K. Cremanns[2], A.-H. Bonardi[2], J. Göttert[2]

Niederrhein University of Applied Sciences, Krefeld, (Germany),
[1]Institute for Coatings and Surface Chemistry (ILOC)
[2]Institute for Surface Technology (HIT)
christian.schmitz@hsnr.de

The development of new recipes for coating material based on customer demands requires a broad overview and knowledge on all raw materials and its properties. Oftentimes, the number of ingredients exceeds 10 or 15 complicating the task of improving product qualities or developing coating properties on demand. The acceleration of the research and development shortens the time to market and amortisation of the costs. High-throughput equipment automates manual lab processes from the dispensing and mixing of the raw materials over the application and hardening of the coat and characterization and quality control of the coated substrate. In comparison to automated production lines, which basically repeat the same procedure, the automation of research requires flexibility treating each sample individually. Thus, every single process step and characterization method being used in the automated workflow should be considered carefully to get promising results for the data analysis. A choice of a smaller number of universal testing methods or simplification of the manufacturing process, which still allow a correlation to the typical lab procedures conducted manually, will very likely improve the applicability.
Statistical methods like design of experiment support the planning and analysis of the data obtained. Modern data analyses allow a detailed look on the system under examination rather than only using ratios and concentration of ingredients or simply comparing the performance of two or more kind of raw materials. More elaborated models make use of next level data analysis linking values of the materials (e.g. hydroxy content of resin, polarity of an additive) to properties of the final product. For this reason the database structure is important making the data accessible for the statistical analysis.
Saving these values of the raw materials or even interactions between raw materials being saved as sub-models into the database simplifies the decision to be made for future improvements or new projects. Connecting results between labs or different projects helps to complement each other achieving a better understanding of the overall activities. This promise a digitalization of research and development based on algorithms, but comparability and introduction of standardization for all participators involved are absolutely necessary.

Light-curing varnish with reduced flammability dedicated to wooden and wood-based substrates

Bartosz Kopyciński, Ewa Langer, Marzena Nowicka-Nowak, Sebastian Jurczyk, Izabela Gajlewicz, Marta Lenartowicz-Klik

Łukasiewicz Research Network – Institute for Engineering of Polymer Materials and Dyes, Toruń, (Poland), e-mail: bartosz.kopycinski@impib.lukasiewicz.gov.pl

Despite the growing awareness of fire protection, over 36,000 fires in residential and industrial buildings were registered in Poland only in 2021 [1]. This creates the need to develop new coating solutions, especially with a more environmentally friendly profile. These efforts include, among others, resignation from the use of halogen flame retardants, which, when heated, can decompose into the toxic dioxins that can accumulate in the human body [2], and formulation of varnishes without the use of organic solvents [3].

A flame-retardant, UV-cured coating composition dedicated to wooden and wood-based substrates was prepared. Based on a mixture of an aliphatic urethane acrylate with a polyester resin, a varnish containing an environmentally friendly, halogen-free flame retardant system as an ammonium polyphosphate and pentaerythritol was obtained. The characteristics of selected physicochemical properties of compositions with variable proportions of flame retardants were done. The hardness of the tested coatings was determined by the pendulum damping method and adhesion to substrate by cross-cut test. Thermal analysis was performed using differential scanning calorimetry and thermogravimetry. The material's flame retardancy was assessed using a cone calorimeter by analysis of the burning process.

The best set of physicochemical properties was obtained with the use of 30 wt% flame retardants addition into the polymer matrix. An increase in the temperature of decomposition T_c by nearly 24°C was observed without any significant influence on the polymerization process. The hardness has increased by nearly 30 seconds. The cone calorimeter analysis showed a 60% reduction in the heat release rate HRR and a 30% reduction in the total smoke production TSP compared to the unmodified material. The cross-linked coating composition showed good resistance to delamination from wooden and wood-based substrates (damage ≤15%).

It is anticipated that the use of the abovementioned varnish may contribute to both increasing the safety of use and extending the lifetime of the covered wooden and wood-based elements.

References:

[1] https://www.gov.pl/web/kgpsp/statystyczne-podsumowanie-2021 [last accessed 10 January 2022]
[2] Shaw, S. D. et al. Halogenated flame retardants: do the fire safety benefits justify the risks? Reviews on Environmental Health. 25 (4), 261-305, (2010)
[3] Shukla, V. et al. Review of basic chemistry of UV-curing technology. Pigment & Resin Technology. 33 (5), 272-279, (2004)

Application of Nuvolve ^{TM} Engineered PolySaccharides in Coatings

Jarmila Vlasáková[1], Richard Milič[1], Kyle Kim[2]

1 - SYNPO a.s., Pardubice (Czech Republic), jarmila.vlasakova@synpo.cz
2 - DuPont BioMaterials (USA)

Advances in industrial coatings have focused on providing improved paint performance while also optimizing aspects such as pigment efficiency (e.g. reduction of TiO_2), gloss reduction while maintaining key properties such as abrasion performance and overall mechanical properties. At the same time, continued emphasis has been placed on reducing the environmental footprint through lower volatile organic content in paint systems. In addition, increasing efforts have been directed to introduce and eventually replace typical building blocks in coatings formulations with more sustainable material alternatives.
Enzymatic polymerization is advancing as a new industrial process option to provide engineered polysaccharide materials with unique structural polymer properties as well as differentiated material morphology.
Our work is focused on using of the alpha-1,3-glucan polymer. The unique microstructure of the polysaccharide allows it to be easily formulated into various paint resins with excellent compatibility. The first part of this work is focused on the improvement of properties of common commercially relevant coating systems by selected engineered polysaccharide (Nuvolve^{TM}, glucan) dispersions. The second part of work contains evaluation of alkyd and acrylic water borne paints, where TiO_2 was partially replaced by the glucan dispersions. Typical standard coating assessments was carried out to assess the performance. Evaluation of drying times, hardness development, rheological performance and gloss was done at all paints with or without the glucan additive in the dispersion.
Alpha-1,3-glucan works as a multi-functional additive in these formulations, and when re-dispersed under high shear the polysaccharide is able to increase the hardness of the paint and to reduce the overall dry time. Of special interest is the ability to function as additive that imparts gloss reduction for matting applications in these paint systems.

References:

[1] M. Remaud-Simeon, C. Albenne, G. Joucia, E. Fabre, S. Bozonnet, S. Pizzut, P. Escalier, G. Potocki-Veronese, P. Monsan: Issue Oligosaccharides in Food and Agriculture, *ACS Symposium Series,* **2003,** *849,* 90-103 (2003)
[2] J. van Haveren, E. A. Oostveen, F. Micciche`, B. A. J. Noordover, C. E. Koning, R. A. T. M. van Benthem, A. E. Frissen, J. G. J. Weijnen: Resins and additives for powder coatings and alkyd paints, based on renewable resources, *J. Coat. Technol. Res.*, **2007,** *4* ,177-186, (2007)
[3] Ch. Hans-Dieter: Novel matting agent for low gloss, *Polymers Paint Colour Journal,* **2010,** 200 (4554), 14, (2010).
[4] J. Nasadova: Driers suitable for high solids alkyd paints II, *Conference Proceedings of 44th International Conference on Coatings Technology*, **2013,** 83-90 (2013)

Organofunctional polysiloxanes as precursors of hydrophilic materials

M. Kaczmarek[1,2], *I. Dąbek*[2], *A. Szymańska*[2], *H. Maciejewski*[1,2*]

[1]*Adam Mickiewicz University, Poznań, (Poland),*
e-mail: marta.kaczmarek@amu.edu.pl
[2]*Poznań Science and Technology Park, Adam Mickiewicz University Foundation, Poznań, (Poland)*

Along with the progress in science and the development of industry, an increasing interest in materials with defined surface properties is observed. One way to define such properties is to determine how they interact with various liquids, including water. If the surface is well wetted by water, meaning water spreads evenly over it without creating droplets, then that surface is called hydrophilic [1]. Due to their interesting self-cleaning and anti-fogging properties, such surfaces are used in the production of windows and automobile glasses, mirrors as well as in biomedical and filtration applications [2,3]. Determination of hydrophilic character of the materials is possible by measuring the water contact angle. The surface can be called hydrophilic when the water contact angle is smaller than 90° [2]. Hydrophilic surfaces can be obtained by two main methods: modification of surface chemistry or by coating the surface of materials with hydrophilic compounds. In the latter method, application of organosilicon compounds containing polar groups such as, e.g. hydroxyl, carboxyl or sulphonic allows to obtain hydrophilic materials [1,3].

In this communication, a method of synthesis of new, hydrophilic organosilicon compounds containing various functional groups and their spectroscopic characteristics will be presented.

Acknowledgments:

This work was supported by the National Science Centre Grant OPUS UMO-2018/29/B/ST8/00913.

References:

[1] Arkles, B. Hydrophobicity, Hydrophilicity and Silanes. *Paint and Coatings Industry*. 22 (10), 114-135, (2006)

[2] Ahmad, D., van den Boogaert, I., Miller, J., Presswell, R., Jouhara, H. Hydrophilic and hydrophobic materials and their applications. *Energy Sources, Part A: Recovery, Utilization, and Environmental Effects.* 40 (22), 2686-2725, (2018)

[3] Sethi, S. K., Manik, G. Recent Progress in Super Hydrophobic/Hydrophilic Self-Cleaning Surfaces for Various Industrial Applications: A Review. *Polymer - Plastics Technology and Engineering.* 57 (12), 1-21, (2018)

New Type II Photoinitiator System for Free Radical Polymerization for Coatings

T. Poplata, *H.Hartmann, B. Strehmel*

Hochschule Niederrhein, Krefeld, Germany, taner.poplata@hs-niederrhein.de

Technische Universität Dresden, Dresden, Germany, horst.hartmann@tu-dresden.de

We disclose in this contribution a new photonitiating system for free radical polymerization. It comprises the thiophene based structure I and an iodonium salt of the the general structure II. Photoinduced electron transfer results in bleaching of I. The polymer films obtained were characterized with respect to their mechanical and thermal properties. Multifunctional monomers of structure III operated as matrix materials to solidify the matrix following a chemical drying mechanism.

The samples were irradiated with a 395 nm LED and the double bond conversion was determined by real-time FTIR. The measurements showed that conversions of up to 80% could be achieved which exhibits a great thing in this field. Such turnovers in relation to a high reaction rate indicate a photoinitiator system with high potential that will be introduced here for the first time at this conference series.

The effect of bleaching is of great advantage since significantly better results can also be achieved in terms of deep curing. With regard to the crosslinking of the systems, results were also obtained which speak in favor of using the compounds disclosed as sensitizers to initiate radical photopolymerization following a free radical photopolymerization. Double bond conversion of up to 85% could be achieved.

In the next step, further investigations were carried out to characterize the derivatives more precisely. For the tests, renewed methyl methacrylates were used as binders in combination with the photoinitiator system consisting of sensitizer and iodonium salt. The reaction kinetics of the respective sensitizers were to be investigated. This was done by means of photo-DSC.

As the synthesis of binders and especially with respect to tailor-made systems is becoming more and more important, we are currently focusing on the use of derivatives disclosed in a metal-free photo-ATRP with UV or blue light. The metal-free photo-ATRP experiment is performed by means of a photoreactor comprising of four 405 nm LED sources and the use of MMA as well as ethyl a-bromoisobutyrate in combination with the selected photoinitiator system. The use of the derivatives disclosed here should be of great advantage within the metal-free photo-ATRP as shown first time in this contribution.

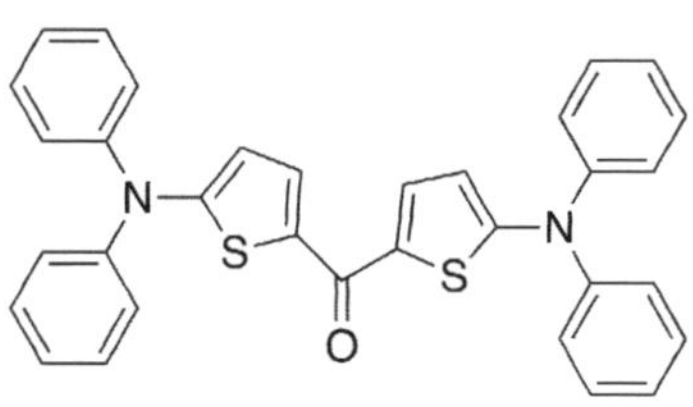

CH-10 I

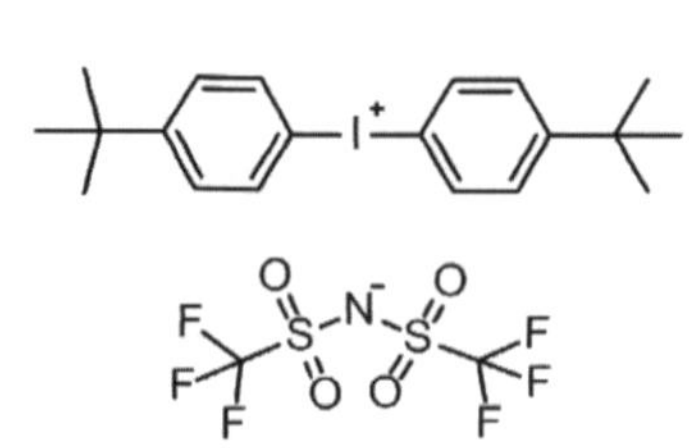

II

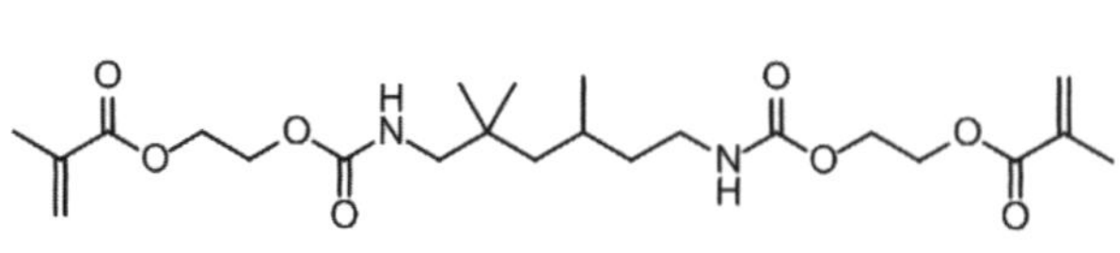

III

The effect of thermomechanical processing parameters on selected properties of thermoplastic starch

A. Krasuska, A. Pietruszka, A. Wojtala, W. Janik

Łukasiewicz Research Network – Institute of Heavy Organic Synthesis "Blachownia", Kędzierzyn-Koźle, (Poland), agata.krasuska@icso.lukasiewicz.gov.pl

Due to the increasing demand for petroleum materials and the impact of their accumulation in the environment, there is a growing interest in biodegradable and compostable polymers development [1]. Among a wide range of biopolymers, starch is the most popular natural polymer. It results from its availability, low cost and easy degradation. Solution casting is the most traditional method used to obtain starch coatings. This method is suitable for producing materials with different thicknesses for food packaging purposes [2, 3]. On the other hand, the method has some disadvantages, such as high energy inputs requiring and low productive. The thermomechanical processing is another method of the starch coatings development . The so-called thermoplastic starch, which is obtained by thermomechanical plasticization of starch with water or glycerol, is becoming increasingly popular [4]. The subject of our studies was the starch plasticization process carried out in the presence of glycerol in a batch mixer under variable conditions (e.g. temperature, time and rotational speed of mixer rotors). The effect of processing conditions on the value of melt flow index and tensile strength of the material was evaluated. Flexible, transparent moldings were obtained regardless of processing conditions. It was discovered that the temperature and processing time had relevant influence on the values of melt flow index. On the other hand, these parameters had lower impact on the mechanical properties of the obtained material – tensile strength and elongation at break. It was also found that mechanical properties of tested samples did not change significantly during storage for 20 months, which proved the stability of the material.

References:

[1] Moshood T. D., Nawanir G., Mahmud F., Ahmad M. H., Ghani A. A. Biodegradable plastic applications towards sustainability: A recent innovations in the green product. Cleaner Engineering and Technology. 6 (2022) 100404

[2] Chi K., Wang W., Catchmark J. M. Sustainable starch-based barrier coatings for packaging applications. Food Hydrocolloids. 103 (2020) 105696

[3] Pelissari F. M., Ferreira D. C., Louzada L. B., Fabiana dos Santos, Corrêa A. C., Kley F., Moreira V., Mattoso L. H. Starch-based edible films and coatings: An eco-friendly alternative for food packaging. Starches for food application. Chapter 10. ISBN 978-0-12-809440-2. 2019

[4] Khan B., Niazi M. B. K., Samin G., Jahan Z. Thermoplastic Starch: A Possible Biodegradable Food Packaging Material—A Review. Journal of Food Processing Engineering. 40 (2017) e12477

The influence of polyurethane roof insulation thermal aging in air and seawater on structural, physico-chemical and mechanical properties

I. Barszczewska-Rybarek, K. Jaszcz, G. Chladek, P. Grabowska, A. Okseniuk, M. Szpot, M. Zawadzka, A. Sokołowska, A. Tarkiewicz

Silesian University of Technology, Gliwice (Poland),
e-mail: Izabela.Barszczewska-Rybarek@polsl.pl

In recent years, the discussion on problems of modern building construction is dominated by issues related to the increasing requirements of energy efficiency. The rigid polyurethane foam (PU) offers the best solution for these problems. If compared to other popular thermal insulating materials, such as expanded polystyrene and mineral wool, it offers a splendid thermal insulation performance. It has the lowest thermal conductivity, which reflects in the lowest insulation thickness and, consequently, a significant reduction of the weight put on the roof. The rigid PU foam is very durable, difficult to deform, has low water absorption and it is resistant to a wide spectrum of chemicals. Additionally, the spray PU foam application process reduces to a minimum the impact of roof thermal bridges and it is easy, quick, free from harmful dusts and fibres.

The purpose of this study was the identification of qualitative and quantitative changes in the visual appearance, chemical structure, morphology and selected physico-mechanical properties of PU rigid foam while heated in air and seawater for predicting its performance and durability. Specimens of the closed cell MDI-based PU foam were aged at constant temperatures of 40 and 80°C for a period of 10 weeks. Annealing caused a yellowing of the samples. Scanning electron microscopy revealed that foam pores lost their regularity. Polymer crumbs were observed at the bottom of cells. Differential scanning calorimetry showed that the glass transition temperature (T_g) increased from 60°C to a maximum value of 77 and 114 °C, respectively for samples stored at 40 and 80°C. It can be explained by the post-polymerization process, which usually results in polymerization shrinkage. The latter one was confirmed by dimensional changes of the studied samples. Their diameter and thickness decreased over time, where the exponential relationship was observed. Water sorption *(WS)* increased over time by around 15 and 25%, respectively for samples stored at 40 and 80°C. These increases can be attributed to the partial cell opening in aged samples. The flexural strength (σ_{flex}) showed a decreasing tendency, whereas for tensile (σ_{tens}) and compressive (σ_{comp}) strengths an increasing tendency was observed. The starting material was characterized by the σ_{flex}, σ_{tens} and σ_{comp} of: 436, 413 and 284 kPa, respectively. The lowest determined σ_{flex} was of 345 kPa. The highest determined σ_{tens} and σ_{comp} equalled 608 kPa and 323 kPa, respectively. The changes in the modulus of elasticity were statistically insignificant and its average value was of 4.0 MPa. The scale of all changes was greater at 80°C.

The study was financed by POWR-03.05.00-00-Z098/17/00 "*Silesian University of Technology as a Center of Modern Education based on research and innovation*", project co-financed from the EU funds and by own funds of the Faculty of Chemistry, Silesian University of Technology.

Bio-based additives for coil coatings

Vladimir Levchenko, Gabriela Gil Alvaradejo, Daniel Schaupp, Charlotte Herstad

Dynea AS, Lillestrøm, Norway, vladimir.levchenko@dynea.com

Growing environmental concerns have increased the interest in the synthesis of sustainably produced polymers. This aims to reduce the dependence on fossil fuels that would result in positive environmental impacts such as reduced carbon dioxide emissions.[1] To shift to sustainable alternatives successfully, price and performance are parameters that should be considered. Cost-effective polymers should be produced, that can compete with current oil-based polymers, while at the same time keeping the expected performance.[2]

We designed and synthesized bio-based polymers by incorporating natural-based raw materials with polymerizable groups, such as acrylates and methacrylates. These polymers were then used as additives for coil coating applications and possessed a range of advantages:

- Low foam formation
- Excellent flow and wetting properties
- Performance competitive to fossil-based products

References:

[1] Babu, R.P., O'Connor, K. & Seeram, R. Current progress on bio-based polymers and their future trends. *Prog Biomater* **2,** 8 (2013).

[2] Harmsen, P.F.H., Hackmann, M.M. & Bos H.L. Green building blocks for bio-based plastics. *Biofuels, Bioprod. Bioref.* **8**:306–324 (2014).

Experimental research on a hydrophobic, self-regenerating coating for protection of the surface of painted vehicles

Karol Jędrasiak

The University of Dąbrowa Górnicza (Akademia WSB), Dąbrowa Górnicza, Poland

The subject of the research was to investigate the experimental properties of a hydrophobic, self-regenerating coating for protection of the surface of painted vehicles with the market name POTION. The coating was developed and made available for testing by the Polish company Car Essence. POTION is a polymer coating based on silicone and silicon derivatives. The coating solution was developed to protect primarily commercial motor vehicles, therefore its important features are the lack of the need for product layering, ease of application and, above all, high hydrophobicity and self-regenerating properties that extend the product's service life and prevent the formation of micro scratches. Consequently, the POTION coating is based on only one component, which reduces application time. The effect of hardening and cross-linking of the coating lasts up to 12 hours at room temperature, so it does not require specialized equipment, such as heating lamps or control of room temperature conditions. Based on internal tests lasting over 24 months, the coating manufacturer estimated that the coating lasts up to 36 months and guarantees hardness on a 9H pencil scale.

As part of the study of the coating, 40 different motor vehicles were used. They were all protected with the POTION coating that was the subject of the study. For all vehicles, the hardening time of the coating was up to 12 hours. During typical operation, motor vehicles are exposed to various atmospheric and physical conditions as a result of interaction with the road and particles emanating from the wheels of the preceding vehicles. As part of the tests, the influence of the following factors on the coating was verified: humidity (anti-corrosive properties), resistance to agents used for cleaning vehicles on non-contact car washes, self-regenerating properties, hydrophobic properties, resistance to temperature, visual effects, impact of road salt, exposure to UV radiation , resistance to dirt caused by, among others: insects, metal filings, etc. The test period covered 12 months, in addition, the frequency of washing and the number of kilometres driven were also verified for each vehicle.

The obtained test results confirm that the POTION coating seals the varnish against moisture and is characterized by considerable resistance to cleaning agents with a pH in the range from 4 to 12. It has been experimentally confirmed that the coating has light self-regenerating properties due to high temperature. The coating has strong hydrophobic properties, preventing the formation of so-called waterspots. The coating is transparent with light-diffusing characteristics, thanks to which the car's paint coat becomes more coherent and shinier after application, which translates into the effect of permanent wax. During the study period there were no undesirable side effects such as dulling, fading of varnish.

References:

[1] Ganesh, V. Anand, et al. "A review on self-cleaning coatings." Journal of Materials Chemistry 21.41, 16304-16322, 2011.

[2] Producer of POTION Coating, http://caressence.pl/, 2020.

Surface chemical modification of commercial catheters *via* SI-ATRP

A.Mielańczyk[1], K.Goc[2], O. Klymenko[3], M. Kupczak[1], M. Krasowska[1], D. Neugebauer[1]

[1]*Department of Physical Chemistry and Technology of Polymers, Faculty of Chemistry, Silesian University of Technology, Gliwice, (Poland), e-mail:* anna.mielanczyk@polsl.pl

[2]*Łukasiewicz Research Network — Institute of Non-Ferrous Metals, Department of Hydrometallurgy, Gliwice, (Poland)*

[3] *Department of Histology and Cell Pathology, Faculty of Medical Sciences in Zabrze, Medical University of Silesia, Katowice, (Poland)*

One of the most commonly used materials in the production of catheters are polyurethanes (PU), which are characterized by strength, flexibility, elasticity and lack of susceptibility to thrombosis [1]. Unfortunately, polyurethanes can show allergic or even toxic properties. In addition, they show the low biocompatibility of a hydrophobic surface. However, the last feature can be reduced or even changed by using various types of polyurethane surface modifications [2]. Nowadays, most of the catheters present on the market are physically coated with hydrophilic or bacteriostatic compounds which can be removed easily by merely immersing the catheter in water for 15 minutes. Thus, chemical modification of catheters can be performed, which results in obtaining the desired properties. Controlled polymerization mechanisms such as the surface-initiated ATRP (SI-ATRP) allow producing surface-tethered coatings of controlled thickness, composition, and macromolecular architecture [3].
In our studies, the surface of commercial polyurethane catheters was modified with poly[poly(ethylene glycol) methacrylate] (PEGMA, M_n = 360 g/mol) using the three-step procedure. First, the PU samples were modified with oxygen plasma in order to introduce reactive hydroxyl groups which were subsequently esterified with α-bromoisobutyryl bromide. Finally, SI-ATRP was performed. The chemical composition of the material's surface at each step was investigated by ATR-FTIR spectroscopy. Subsequently, the roughness parameter was determined at individual stages of the surface modification by means of the atomic force microscope (AFM) and scanning electron microscope (SEM). These experiments showed that after the modification (oxidation) process PU became more rough (Ra = 261.50 ± 0.71 nm, Rq = 334.50 ± 0.71 nm) compared to the starting material (Ra = 234.33 ± 83.03 nm, Rq = 328.00 ± 139.15 nm). After esterification, microfissures were observed in the material. Finally, the lowest values were observed in the PU-*graft*-P(PEGMA) material roughness (Ra = 201.67 ± 66.15 nm, Rq = 252.67 ± 83.08 nm). Moreover, the increase in PU surface thickness indicated that the modification was successful.

References:

[1] Wildgruber M., Lueg C., Borgmeyer S., Karimov I., Braun U., Kiechle M., Meier R., Koehler M., Ettl J., Berger H.Polyurethane versus silicone catheters for central venousport devices implanted at the forearm. *Eur. J. Cancer* 59, 113-124 (2016)

[2] Adipurnama I., Yang M-Ch., Ciach T., Butruk-Raszeja B. Surface modification and endothelialization of polyurethane for vascular tissue engineering applications: a review. *Biomater. Sci.,* 5, 22–37 (2017)

[3] Barbey R., Lavanant L., Paripovic D., Schuwer N., Sugnaux C., Tugulu S., Klok H. A., Polymer brushes via surface-initiated controlled radical polymerization: synthesis, characterization, properties, and applications. *Chem. Rev.* 109 (11), 5437- 5527 (2009)

Development of water cycles for sewage treatment of dye-finishing production

A.U. Issayeva, M. G. Koval**, H.S.Stolyarenko ***

** M. Auezov South Kazakhstan State University, Shymkent, Kazakhstan, akmaral.issayeva@bk.ru*
*** Cherkasy State Technological University, Cherkasy, Ukraine radikal@ukr.net*

The results of research on the treatment of industrial wastewater from dyeing and finishing production using filtration, bio-coagulation in cavitators, flotation and electroactivation to develop an effective technology of water circulation cycles and the reuse of industrial wastewater in textile manufacturing processes are presented. The wastewater research results were obtained at an experimental cascade plant, the stages of which were located above the averager - settler of the existing treatment facilities of the enterprise.

It is known that most of the existing dye-finishing production are based on outdated and not perfect dyeing technologies, the wastewater of which is a complex polydisperse systems consisting of ions, dye molecules and auxiliary organic substances. Depending on the type and properties of the textile dye used, the method of dyeing the fabric, from the end of the production cycle 20 to 50% of the dyes fall into the wastewater, which leads to significant damage to the ecological system of the region where such productions are located. The existing dye-finishing industries can`t be called effective and highly profitable, because the cost of dyeing fabric to a large extent depends on the cost of dyes and the degree of their use.

Development purpose: Creation of a new concept of deep wastewater treatment containing organic dyes, study of the possibility of reuse of the dye components that make up the wastewater in the technological cycle of dyeing and finishing production.

Two types of wastewater were examined: industrial wastewater immediately after the fabric dyeing process (at the exit from a dyeing machine with automatic Super-Jigger VN machines by «Henriksen», Denmark), as well as wastewater after averaging and dilution, that is, the waters of the enterprise, which according to the current technology is sent to the city treatment system. A local water purification system based on filtration and electroactivation processes makes it possible to increase the percentage of dye use from 46-50% to 68% when reusing water in the fabric dyeing cycle. Wastewater of the enterprise with dyes of various origins enters the averager, undergoes a filtration process on a recycle filter from sawdust. Spent sawdust is either burned or composted. Next, water with fine particles of dyes is subjected to treatment with a biocoagulant in cavitators and enters the sedimentation. Correction for pH and final dye concentrations is carried out using unipolar electrochemical activation. Its purpose is to achieve dye content that does not affect the quality of secondary dyeing. Water is partially or fully returned to the enterprise. According to the control, the following wastewater treatment plant efficiency was achieved: COD decreased by 32 - 48%; BOD - by 26 - 38,6%, the content of dyes - by 65,6 - 88,4%. Cleaning processes lead to a fairly good discoloration (90%), an increase in transparency (5 ... 7 times), and a decrease in the content of suspended solids (60%).

Reuse of water minimizes operating costs and solves the problem of creating a recycled water supply system for the enterprise, and also leads to an increase in its economic efficiency and high profitability.

SILP materials as new type initiators of the epoxy resins - from synthesis to applications

Dawid Zielinski[1,2], *Andrea Szpecht*[1], *Rafal Kukawka*[3], *Paulina Nadobna*[1], *Joanna Dzialkowska*[2], *Mariusz Pietrowski*[2], *Michal Zielinski*[2], *Marcin Smiglak*[1,3]

[1]Poznan Science and Technology Park, Poznan, Poland
[2]Adam Mickiewicz University, Poznan, Poland
[3]Innosil Sp. z o.o., Poznan, Poland
Email: dawid.zielinski@ppnt.poznan.pl

The main goals of the research are developing the innovative materials technology in the field of novel, multi-functional systems of epoxy composites, containing ionic liquids (ILs) as curing agents. ILs in this studies was used as SILP materials (Supported Ionic Liquid Phase). It is necessary to research for new composite technologies to make production processes more sustainable and greener. In this research, innovative, based on SILP materials as hardeners and commercial epoxy resins systems were designed. One of the most interesting property of ILs is their ability to initiate polymerization of epoxies [1]. In detail, these compounds act in epoxy systems in a role of latent curing agents, which need thermal energy for activation. In contrast, SILP is a material in which the ionic liquid is deposited on a solid support, e.g. silica [2]. Such a system protects the ionic liquid bound to the silica surface as well as remains in solid form - which allows it to be used as a composite filling. SILP can act as a hardener and filler in the epoxy system, and the ability to design it (matching the amount and type of ionic liquid deposited on the selected type of silica) allows to accurately develop an epoxy formulation for a specific application. New SILP materials initiating polymerization, presented in the paper, have been extensively characterized in terms of determining their specific surface area, pore size and degree of saturation. Physicochemical properties were enriched with infrared spectroscopic spectra (FT-IR). In addition, the materials obtained were imaged using Scanning Electron Microscopy (SEM) with mechanical analysis. Studies on the application potential and usability in the production of epoxy composites have been carried out on systems with commercial epoxy resins from various manufacturers.

This work was supported by the National Science Centre (Poland), project SONATA BIS no UMO-2017/26/E/ST8/01059.

References:

1. Rahmathullah, M. A. M. et al. Room temperature ionic liquids as thermally latent initiators for polymerization of epoxy resins. Macromolecules 42, 3219–3221 (2009).
2. Kukawka, R. et al. Highly Effective Supported Ionic Liquid-Phase (SILP) Catalysts: Characterization and Application to the Hydrosilylation Reaction. ACS Sustain. Chem. Eng. 7, 4699–4706 (2019).

Ruthenium Rigidochromic Dyes as Probes for Resin Curing

E Savino, L E Eijsink, A S Sardjan, E G Sinnema,W R Browne

Photocurable resins have diverse applications including additive manufacturing, coatings, dental medicine, and prosthetics.(1) Understanding curing kinetics and especially the time to reach the vitrification point is important to these applications, however the very high viscosities limit the number of methods applicable to studying such systems. Indeed for photocuring, calorimetry is the method of choice in terms of analysis of rate and extent of curing(2,3). In this presentation, spectroscopic methods for analysing photocuring is described. Micro-Raman and time resolved UV-Vis absorption spectroscopy is used to monitor the change in the chemical composition of the photosets and is combined with EPR and emission spectroscopy (in presence of the rigidochromic complex $[Ru(dpp)_3](PF_6)_2$) under irradiation to relate the activation of radical initiators with the viscosity of the resins. This approach is evaluated on custom made and commercially available photocuring resins.

Bibliography:

1) **https://doi.org/10.1002/admt.201800618**
2) **https://doi.org/10.1002/pola.20529**
3) https://doi.org/10.1007/s10973-014-3984-z

Nanosized MoS_2 deposited on carbon nanomaterials as the new lubricant additive for engine oils

Zuzanna Bojarska, Radosław Lewandowski, Janusz Kopytowski, Antoni Rożeń, Łukasz Makowski

Faculty of Chemical and Process Engineering, Warsaw University of Technology, Warsaw (Poland), zuzanna.bojarska.dokt@pw.edu.pl

The development of new advanced lubricant additives is crucial for the production of cleaner and more durable internal combustion engines. The tribological and rheological properties of engine oils have a significant impact on engine efficiency. The additives are used in order to reduce friction resistance, protect moving engine elements against wear, improve fuel economy and reduce pollution of gases emitted to the environment. Over the last few years, various types of nanoparticles have been tested for additives in engine oils. Molybdenum disulphide nanoparticles (MoS_2) deposited on carbon nanomaterials (CNMs), such as graphene oxide, seem to be very promising material, due to their synergetic lubricating effect [1,2].
The aim of this study was to develop a method of the facile and scalable method to produce MoS_2/CNMs hybrid nanocomposites in micromixer. Subsequently, the obtained nanocomposite has been investigated as the additives for engine oil (Liqui Moly 10w40). The rheological properties have been measured in rheometer 302 from Anton Paar. Different ratios of the nanosuspensions were tested (1 wt.%, 5 wt.% and 10 wt.%) at different temperatures (-10 °C, 0 °C, 25 °C, 75 °C). The tribological properties of the same nanosuspensions have been tested in MCR Tribology Cell T-PTD 200 with the ball-on-three-plates setup at the same temperatures. Furthermore, the obtained results were compared with the tribological and rheological date of pure synthesized MoS_2 nanoparticles obtained by wet chemical synthesis and with commercial MoS_2 particles provided by Sigma- Aldrich (<100 nm). Chemical composition and structure of the obtained materials were analysed using scanning electron microscopy, thermogravimetric analysis, and particle size analysis by laser diffraction.
The wet chemical synthesis in the micromixer is a proper and facile approach to produce the hybrid nanocomposites. The obtained results show that the addition of molybdenum disulphide and materials based on MoS_2 to engine oil reduced its friction coefficient and viscosity. The suspension with the addition of the hybrid nanocomposite showed the best tribological properties. The addition of carbon materials to MoS_2 reduces the agglomeration of particles, thereby causing less sedimentation of nanosuspensions. The obtained materials may compete with existing commercially available oil additives.

Acknowledgments:
This work was supported by the National Science Centre [No.2017/27/B/ST8/01382]

References:

[1] Song, H., Wang, B., Zhou, Q., Xiao, J., Jia, X. Preparation and tribological properties of MoS_2/graphene oxide composites. *Applied Surface Science*. 419, 24 – 34, (2017)
[2] Chen, Z., Yan, H., Liu, T., Niu, S. Nanosheets of MoS2 and reduced graphene oxide as hybrid fillers improved the mechanical and tribological properties of bismaleimide composites. *Composites Science and Technology*. 125, 47 – 54, (2016)

Effect of pigments and fillers on properties of new acrylate dispersion containing nanoparticles MgO

D. Steinerová, A. Kalendová, J. Machotová

Institute of chemistry and technology of macromolecular materials, Faculty of chemical technology, University of Pardubice, Studentská 573, 532 10 Pardubice

ABSTRACT

This paper deals with characterization and study of environmentally friendly anticorrosive paints, based on newly synthesized acrylate water-dilutable polymer dispersion with presence of 1.5% magnesium oxide nanoparticles (relative to monomers) prepared by semi-continuous emulsion radical polymerization. Synthesized binder shows very promising mechanical, chemical and antimicrobial properties, but low corrosion resistance. For this reason, the formulation was performed with environmentally friendly pigments of different chemical composition and particle shape and their effect on the anticorrosive, mechanical and chemical properties of the coating film was monitored. The evaluated results showed higher protective properties of coating films containing electrochemically acting pigments.

Novel Polyamide Rheology Control Additives for Improved Orientation of Metallic Pigments

D.Oprych[a)], M. Ubukata[b)], H. Yamada[b)], T. Ikuta[b)]

[a)]C.H. Erbslöh GmbH & Co. KG, 47809 Krefeld, Germany
[b)]Kusumoto Chemicals Ltd., 101-0047 Tokyo, Japan

Amide compounds used as rheology control additives (RCAs) for solvent-borne and solvent-free systems are wax-like materials with a high melting point due to their diamide structure and are called amide waxes or polyamides. Amide wax has a needle-like crystal structure with a three-dimensional network structure in the system that imparts pseudoplasticity. To develop rheology control, amide waxes must swell and be finely dispersed through the activation process, leading to the aggregation of acicular crystals. This activation process generally requires heating and shearing. Thanks to its dense crystal structure, well-processed paste-type RCAs do not cause turbidity or adversely affect the smoothness of the coating film and so are used in applications that require a superior appearance (Figure 1). Today, all industries need to address environmental issues to maintain sustainable growth into the future. In the field of coatings, volatile organic compounds (VOC) have been the major problem because of their risk to health and the environment. The metallic paints used in the automotive coating especially require quite a large quantity of organic solvents to achieve the superior appearance by controlling the orientation of metal pigments. Global VOC regulations are getting stricter year after year and the automotive industry has been forced to change its metallic paints from solvent-borne to water-borne without compromising appearance. Amide wax used in solvent-borne and solvent-free systems has a hydrophobic structure, and it is difficult to adapt it to water-borne systems. Therefore, a low molecular weight amide polymer with an anionic hydrophilic group is used as a polyamide rheology control additive (PARCA) for water-borne systems.

We tested a series of RCAs in an acrylic urethane metallic coating system of the type typically used for automotive coatings. The anti-settling properties of these paints are also evaluated. The metallic coatings were applied on polycarbonate panels by air-spray. After drying, we measured the L* values of films using a multi-angle spectrophotometer and calculated the flop indexes to establish the flip-flop. Ethylene-vinyl acetate (EVA) wax was effective with regard to orienting the metallic pigments, but it required a large dosage and even at such a high dosage, EVA wax could not provide the system with sufficient anti-settling. The hybrid RCA with amide wax and oxidized polyethylene wax imparted higher pseudoplasticity than EVA wax and achieved an excellent anti-settling effect and pigment orientation even at low dosage.

(TEM image)

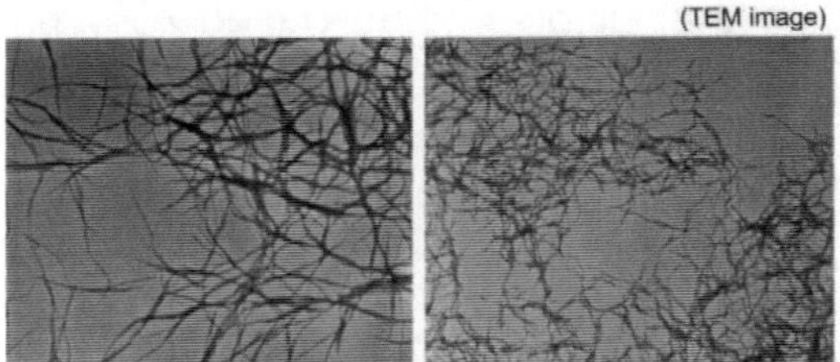

Activated amide particles — Fine amide particles

Fig.1-1 Activated amide particles

6% in an acrylic resin

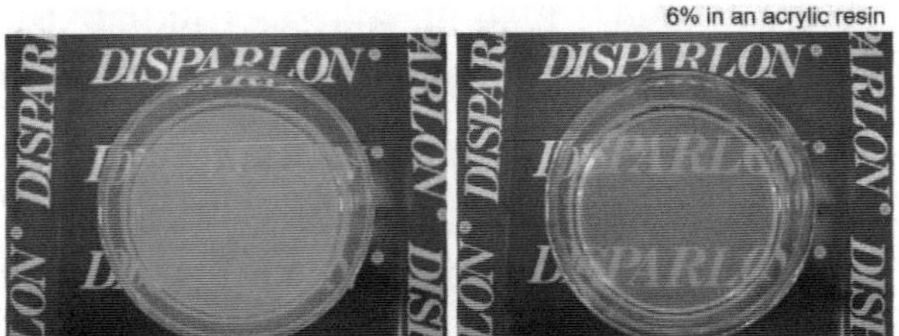

With conventional amide particles — With fine amide particles

Fig.1-2 Transparency

Fig. 1-1: TEM images of activated amide particles; Fig. 1-2: Transparency of 6 wt% conventional amide particles in comparison to fine amide particles in an acrylic resin.

[1] Ubukata, M., Yamada, H., Ikuta, T. Eur. Coat. J., 9, 38-45 (2021)

Highly hydrophobic mineral plaster in ready-to-use form

J. Lisowski, B. Szadkowski, A. Marzec

Institute of Polymer and Dye Technology, Faculty of Chemistry, Lodz University of Technology, 90-537 Lodz, Stefanowskiego 16, Poland, jakub.lisowski@dokt.p.lodz.pl

Elevation coatings have been dominated by organic dispersion-based products for decades now, their durability, water resistance, simplicity of use and color stability making them almost flawless materials [1]. However, modern market faces difficulties for their use. Ecological restrictions limiting use of plastic-based materials and many of supporting substances (biocides, surfactants), and recent shortages in polymer materials make space for research of new alternatives.

Concerning growing interest in ready-to-use products in construction materials industry, as well as increasing ecological awareness , especially considering upcoming ECHA restrictions about microplastic use [2], we created plaster in ready-to-use form (traditionally based on organic polymer dispersions), with entirely mineral binder, using water glass.

The proposed formulation was centered around avoiding using of any organic polymer based materials, not only binder, but additives as well (thickeners, rheology modifiers). In present work, they were replaced with either mineral or bio-degradable alternatives.

High pH value of the product (pH~11) makes it possible to desist from biocidic substances, which application in standard formulations is being more strictly regulated. This novel formulation simplifies the process by omitting constant research for new, not-restricted yet biocides and provides plaster with positive pro-ecological character. This regard in-can and shell biocides. To date, there has not been any evidence about bacteria contamination of plaster in can storage or outside after 3 months of exposition.

In this work, the formulated plaster was prepared to match application standards of polymer dispersion-based plasters, in other words the designed plaster exhibited similar rheological and aesthetic properties. Material was modified using potassium methyl siliconate and classified mix of silanes/siloxanes to achieve very high hydrophobic properties, which are crucial for elevation coatings, showing smaller water uptake than standard silicone modified dispersion plasters. Moreover, it has excellent adhesion to mineral substrates, so it doesn't require use of primers on the walls.

Prototype plaster performed very well in terms of application properties, having workability and substrate adhesion comparable to that of dispersion-based plaster. Excellent water resistance makes it competitor for top polymer plasters, and reducing additives' amount in formulation indicates savings to production costs. These properties make it potentially environmental and wallet friendly alternative to polymer-based ready-to-use products.

References

[1] Sharma, V. Polymers and Microplastics: Implications on Our Environment and Sustainability. Emerging Technologies, Environment and Research for Sustainable Aquaculture. 39-40, (2020).

[2] https://echa.europa.eu/pl/hot-topics/microplastics

Laser diffraction measurements of pigment size distributions in coatings and mill bases

Shicong Luo, Claus Erik Weinell, Søren Kiil

Department of Chemical and Biochemical Engineering, Technical University of Denmark, DTU, Kgs. Lyngby, Denmark, shiluo@kt.dtu.dk

Most performance properties of coatings, such as rheology, colour strength, UV radiation resistance, and gloss, can be related to the degree (or lack) of dispersion [1,2]. When evaluating the presence of agglomerates in coatings or mill bases, the industry generally applies the maximum particle size as the critical parameter. However, the entire particle size distribution (PSD) plays a role in the performance [3]. Over the years, a range of instruments, based on new principles, has been introduced beyond the Hegman gauge [4-6]. However, obstacles are also evident, because these methods are highly influenced by the complex composition and the high solids contents of coatings or mill bases. The aim of the present work was to develop an analysis procedure, based on laser diffraction, for PSD measurements of coating products.

In the four series experiments, acrylic based TiO_2 or Cu_2O mill bases were used, and perturbations in composition and production parameters were investigated. To estimate the PSD of withdrawn samples, a laser diffraction instrument (Malvern MasterSizer 3000, MV chamber) was employed. Sampling from the same batch was done at different dispersion times (i.e. from 15 seconds and up to 40 minutes).

The PSD measurements were influenced by the obscuration value of the sample, the refractive index input values, and the shape of the pigment. Furthermore, prior to a measurement, the mill base (or coating) sample had to be diluted, which affects the stability of particles and agglomerates.

When using pure xylene as the measurement medium, a solvent shock, which resulted in non-representative PSDs, was observed. However, when using a binder-solvent mixture, taking rheological considerations into account, solvent shock was avoided and reliable PSDs could be obtained. Furthermore, to detect the transient PSD and volume-moment mean diameter, associated with variations in coating formulations and operation parameters, the analytical range of the instrument was sufficient.

In summary, the principle of laser diffraction was demonstrated to be a reliable tool for PSD estimations for mill bases and coatings, as long as the dilution process is properly analyzed and controlled.

References:

[1] Winkler, J., Dispersing pigments and fillers, Hannover, Germany, Vincentz, (2012)

[2] Pieri, L., Bittelli, M. Laser diffraction, transmission electron microscopy and image analysis to evaluate a bimodal Gaussian model for particle size distribution in soils. Geoderma, 135, 118-132, (2006)

[3] Brody, D. E. PERFORMANCE OF FLATTING PIGMENTS IN HIGH-SPEED DISPERSION, Journal of paint technology, 40(525), 439, (1968)

[4] Oyarzún, Juan, M., Pigment processing: physico-chemical principles. Vincentz Network, (2000)

[5] Servais, C., Jones, R., & Roberts, I. The influence of particle size distribution on the processing of food. Journal of food engineering, 51(3), 201-208. (2002)

[6] Allen, T. Particle size measurement. Springer. (2013)

Photocuring of Nanoparticle-Comprised Coatings Based on Heptamethine Cyanines with NIR Exposure between 800-950 nm

Qunying Wang, Bernd Strehmel
Department of Chemistry, Institute for Coatings and Surface Chemistry, Niederrhein University of Applied Sciences, Adlerstr. 1, D-47798 Krefeld, Germany

Email: qunying.wang@hs-niederrhien.de

High-intensity emitting NIR-LEDs have attracted great interest among the research for application in photopolymerization particularly for uses in coatings applying chemical drying. It possesses some benefits compared to UV initiated polymerization; that is the use of monomers and additives exhibiting an intrinsic absorption in the UV. The longer wavelength also facilitates deeper penetration of excitation light and offers therefore more possibilities for photocuring. In this work, cyanines derived from heptamethines comprising a special conjugated pattern facilitate both initiation of free radical and cationic photopolymerization as combined with new high-power NIR-LED emitting at 820nm that are available as prototypes in our lab. Iodonium salts with low coordinating anion based on perfluoroaluminates serves as co-initiator – just to name one example.

Herein, we report for curing of epoxides comprising sol-gel forming materials based on functionalized silanes resulting in polymer films exhibiting higher T_g from NIR initiation system than UV as determined by DMA measurements. In addition, the heat generated by the NIR-sensitizer by non-radiative deactivation caused by internal conversion uptakes a special function, which benefits the generation of initiating species such as radicals and conjugate acid for free radical and cationic polymerization, respectively, based on a photonic process.

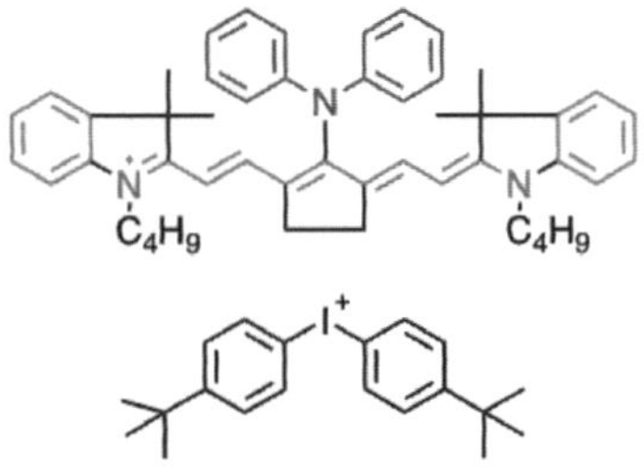

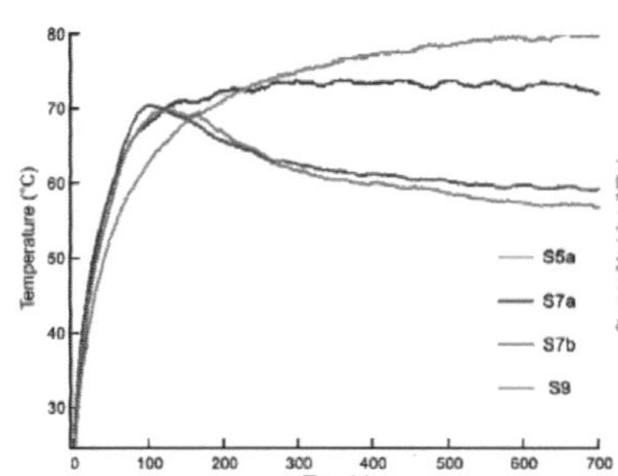

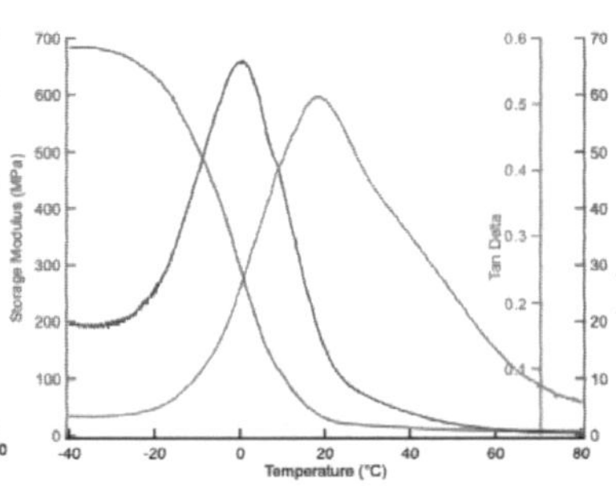

Acknowledgement
We acknowledge Selina Gomoll from FEW Chemicals for providing the raw materials of sol-gels used in this work.

Novel Supported Ionic Liquid Phase (SILP) materials as effective catalysts in organofunctional silanes synthesis

Rafal Kukawka[1], Joanna Dzialkowska[2], Marcin Smiglak[1], Hieronim Maciejewski[1,2]

[1]Poznan Science and Technology Park, Poznan, Poland
[2]Adam Mickiewicz University, Poznan, Poland
Email: rafal.kukawka@ppnt.poznan.pl

Organosilicon compounds, because of their unique properties, are widely used in a variety of organic processes, and thus the constant improvement of current methods is still needed. Synthesis of most of organosilicone compounds is based on hydrosilylation reaction.[1] The major problem is the difficulty in the separation of catalyst from a post-reaction mixture. The presence of metal in the reaction products, even at the trace quantities, is unacceptable for many applications (especially in food and cosmetic industry). Therefore, efforts are made to utilize heterogeneous catalysts. One of the possible solutions to this problem is carrying the reactions in biphasic system with catalyst dissolved in ionic liquid supported on solid support (SILP). The combination of ionic liquids, inorganic supports, and platinum complexes can allow the synthesis of organosilicon compounds in batch and continuous flow systems and may be solution for a variety of problems of today's processes.

The results presented in the work focus on the synthesis on novel SILP and their application as effective catalysts in synthesis of organofunctional silanes via hydrosilylation reaction in batch and continuous flow reactor system using scCO2 as mobile phase. The obtained new SILP catalysts were analyzed by infrared technique, low-temperature nitrogen physisorption at 77 K, and scanning electronic microscopy with energy-dispersive X-ray spectroscopy to provide structural information on these materials. Moreover, the catalytic activity in hydrosilylation reactions was evaluated and compared with the catalytic activity of rhodium catalysts dissolved in the same ILs when using a biphasic reaction system (IL/catalyst as one phase and mixture of substrates as a second phase). The rhodium-based SILP catalysts proved to be much more efficient than when used in a biphasic system composed of a similar catalyst and reactants.

This work was supported by the National Science Centre (Poland), project SONATA BIS no UMO-2017/26/E/ST8/01059.

References:

[1] Cornils, B., Hermmann, W.A., Belier, M., Pawelo, R. (Eds.) *Applied Homogeneous Catalysis with Organometallic Compounds*, John Wiley & Sons, 569–620, (2017).

[2] Fehrmann, R.; Riisager, A.; Haumann, M. (Eds.) *Supported Ionic Liquids: Fundamentals and Applications*; John Wiley & Sons (2014).

[3] Kukawka, R.; Pawlowska-Zygarowicz, A.; Dzialkowska, J.; Pietrowski, M.; Maciejewski, H.; Bica, K.; Smiglak, M. *Highly Effective Supported Ionic Liquid-Phase (SILP) Catalysts: Characterization and Application to the Hydrosilylation Reaction. ACS Sustain. Chem. Eng.* 4699–4706, (2019).

Study of the minimal surface-based solids as an infill of the 3D printed objects and their effect on compressive strength

Petr Resl, *Luboš Prokůpek*

University of Pardubice, Pardubice (Czech Republic), petr.resl@student.upce.cz
University of Pardubice, Pardubice (Czech Republic), lubos.prokupek@upce.cz

As 3D printing rises in its popularity, opening a new opportunity for investigating. By its meaning, the 3D print is the numeric controlled selective additive manufacturing process. The material is always used in a particular amount and layered down to specific coordinates on the top of the previous layer. In addition, 3D print is independent of the shape, therefore able to create objects even multiaxis subtractive manufacturing is unable to. Triply Periodic Minimal Surfaces (TPMS) are good examples of such shapes. Structures based on TPMS can be applied as an infill of any object.
The amount and structure of the infill can be customized. It leads to material, time, and cost savings. The amount of infill decrees the final strength of products. This poster focus on the advanced structures based on Triply Periodic Minimal Surface and their effect on compressive strength.
Minimal surfaces can be explained as random two-dimensional surfaces in a space that no longer reduces its area with any change of its shape (membranes, canvases, interfaces of two environments). This surface is defined by a random close loop, and within these borders, the surface always creates at least one local minimum, therefore it creates curvature. Some of those minimal surfaces have the property to create three-dimensional lattices by their repetition in space. The result is a three-dimensional structure with an infinitesimal thickness of the faces [1]. This fact complicated the realization of those structures. In this case, two approaches were applied to convert TPMS to solid objects – "Network solids" (NS) and "Sheet solids" (SS) [2]. Those structures are suitable as an infill of any object.
Three TPMS were selected and converted in the solids: Network solid and Sheet solid. Selected TPMS for investigation are called Gyroid, Diamond, and Primitive. The compressive strength and modulus were measured and compared with the reference. As a reference was chosen highly used honeycomb infill. TPMS was designed as the finite lattice and tested in compressive strength, and the deformation was visually evaluated. Those testing samples were printed on the FDM/FFF 3D printer Flashforge Dreamer. Thermoplastic terpolymer ASA (acrylonitrile-styrene-acrylate) was chosen as an example of the constructional plastic for the FDM/FFF 3D printing.
As for the result, the conversion of TPMS to solid has an important effect on the strength. Conversion to NS TPMS creates weaker "necks" in their structures, which deform preferably. Gyroid NS TPMS and Diamond NS TPMS show similar values in the strength and modulus but show interesting behavior in deformation, where those necks translate to the vacation and keep the value of the strength constant during deformation. Another interesting deformation curve was observed at Gyroid SS TMPS and Diamond SS TPMS. No maximum in the deformation curve appeared, and the curve was constantly rising during deformation. Conversion to SS TPMS creates a solid that imitates the original infinitesimal surface and does not create any weak parts.

Keywords: TPMS, 3D printing, strength.

[1] KRIVOSHAPKO, S.N, IVANOV, V.N. Encyclopedia of Analytical Surfaces. *Springer.* 427, (2015)

[2] KAPFER S. C., HYDE, S. T. Minimal surface scaffold designs for tissue engineering. *Biomaterials.* 32(29), 6875-6882, (2011)

Hydrophobization of natural fiber surface with organosilicon compounds

A. Przybylska,[1,2] A. Szymańska,[2] M. Przybylak,[2] and H. Maciejewski[1,2*]

[1] *Adam Mickiewicz University, Poznań, (Poland)*
, e-mail: agnieszka.przybylska@amu.edu.pl

[2] *Poznan Science and Technology Park, Adam Mickiewicz University Foundation, Poznań, (Poland)*

In many applications, surface properties play an important role since they determine the wettability of surfaces with different kind of liquids, deposition of various substances, and the ability to separate liquid mixtures that do not mix. In recent years, interest in the research and production of hydrophobic fabric has increased due to their unique features such as self-cleaning, anti-contamination and anti-sticking [1,2]. Natural fibers, despite many advantages, cannot be used for many purposes due to their high hydrophilicity and polarity. Excessive moisture absorption weakens materials made of natural fibers and at the same time reduces their adhesion and promotes the growth of mold, bacteria or fungi [2,3]. Modification of the surface of textiles allows to give them specific, selected properties in order to increase their area of application [4].

In this communication, synthesis of organosilicon compounds by hydrothiolation and hydrosilylation reactions and their use in hydrophobization of cotton fabrics as well as functionalization of a cotton fabric surface via thiol-ene click reaction will be presented.

Acknowledgments:

The project was supported by the National Science Center Grant OPUS UMO-2018/29/B/ST8/00913.

References:

[1] Xu, L., Zhuang, W., Xu, B., Cai, Z. Fabrication of superhydrophobic cotton fabrics by silica hydrosol and hydrophobization. *Applied Surface Science,* 257(13), *5491–5498,* (2011)

[2] Przybylak, M., Maciejewski, H., Dutkiewicz, A., Dąbek, I., Nowicki, M. Fabrication of superhydrophobic cotton fabrics by a simple chemical modification. *Cellulose*, 23(3), 2185–2197, (2016)

[3] Duan, W., Xie, A., Shen, Y., Wang, X., Wang, F., Zhang, Y.,Li, J. Fabrication of Superhydrophobic Cotton Fabrics with UV Protection Based on CeO2Particles. *Industrial & Engineering Chemistry Research,* 50(8), 4441–4445, (2011)

[4] Erasmus, E., Barkhuysen, FA. Superhydrophobic cotton by fluorosilane modification, *Indian Journal of Fibre and Textile Research*, 34(4), 377-379, (2009)

Water-vapour transmission rate: effect of selected parameters on the example of acrylic-styrene coatings

Małgorzata Gnus, *Grażyna Kamińska-Bach*

The Łukasiewicz Research Network - Institute for Engineering of Polymer Materials and Dyes, Toruń, (Poland), e-mail: malgorzata.gnus@impib.lukasiewicz.gov.pl

In this study, a series of free coatings and coatings on supports made of waterborne paint products containing 7 wt.%, 10.5 wt.% and 14 wt.% of acrylic-styrene binder were performed. Three different supports were used: cellulose, silicone-coated and glass support.

The obtained coatings, coatings on supports and raw supports were tested for their water vapour permeability using the wet cup method described in PN-EN ISO 7783:2018-11[1]. Based on the tests carried out, the influence of the binder content in the paint product on the water vapour permeability values obtained was found. The influence of the type of support on the obtained results was also found.

On the basis of the obtained results, it was observed that the value of the water vapour permeability coefficient decreases with the increase of the binder content in the varnish product, and the observed differences are the greater the greater the thickness of the tested coating.

The water vapour transmission through the substrate depends primarily on the affinity of the material to water molecules and decreases with increasing hydrophilicity of the material.

The ability of the coating to transmit water vapour of the coating obtained on the carrier strongly depends on its interaction with the applied substrate, therefore when determining the permeability coefficient the carrier for testing should be selected carefully.

References:

[1] PN-EN ISO 7783:2018-11 Paints and varnishes — Determination of water-vapour transmission properties — Cup method (2018)

Development of Novel Anti-Fouling Paints for Aquaculture Applications.
In vitro biological evaluation.

Danai Prokopiou[1], Nikoleta Kostopoulou[1], Maria-Malvina Stathouraki[2], Sergios Papageorgiou[2], Eleni Efthimiadou[1,2]
[1]Laboratory of Inorganic Chemistry, Department of Chemistry, National and Kapodistrian University of Athens, Zografou GR-15771, Greece.
[2]INN, NCSR "Demokritos", 153 41 Aghia Paraskevi, Attica, Greece

Biofouling is an extensively studied natural phenomenon that occurs in all immersed underwater surfaces including ships, drilling stations, water inlets and aquaculture equipment such as cages, net and ropes. It is defined as the consecutive accumulation of organic molecules, microorganisms, plants and animals leading initially in biofilm formation and finally in attachment of larger aquatic organisms. But why is biofouling a problem? Regarding ships, the resulting increase in weight and surface roughness affects speed and maneuverability and can lead up to a 40% rise in fuel consumption. In marine aquaculture, biofouling causes cage deformation and net occlusion that restricts water and oxygen exchange leading to accumulation of toxic metabolic products and, in turn, increased level of fish stress, disease vulnerability and lower immunity (Amara et al., 2018). The economic impact is remarkable. Oil, gas and shipping industries spend approximately 5.7 billion US$ annually to prevent marine biofouling. A variety of antifouling methods have been implemented to combat biofouling (Fitridge et al., 2012).

The goal of our research is to develop novel, biocompatible, biocide-free and environmentally friendly water based anti-fouling paints. Aquaculture equipment should be coated with anti-fouling agents before immersing in seawater to prevent or significantly delay biofouling. Our approach includes the development of hybrid paints based on dendritic polymer / silica nanoparticles and/or inorganic nanotubes. Our team is focused on physicochemical characterization and biological evaluation of these nanocomposite paints. Initially, DLS and FT-IR experiments were performed and subsequently, IC50 values and *in vitro* cytotoxicity were evaluated using different human cell lines.

Acknowledgments:
This research was supported by the project "Development of innovative non biocidal antifouling paints for aquaculture applications - AQUAPAINTS" MIS 5030665 funded by the Managing Authority of the Fisheries and Maritime Operational Program of the Greek

References:

Amara, I., Miled, W., Slama, R. Ben, & Ladhari, N. (2018). Antifouling processes and toxicity effects of antifouling paints on marine environment. A review. Environmental Toxicology and Pharmacology, 57(October 2017), 115–130.

Fitridge, I., Dempster, T., Guenther, J., & de Nys, R. (2012). The impact and control of biofouling in marine aquaculture: A review. Biofouling, 28(7), 649–669.

Waterbased, environmental friendly zinc primers

I. Gajecka[*], *U. Paszek*[*], *S. Piłat*[*], *L. Komorowski*[**], *A. Królikowska*[**], *I. Kunce*[**], *M. Urbański*[**], *D. Wojda*[**], *K. Zacharuk*[**], *M. Zubielewicz*[***], *E. Langer*[***]

[*]*Polish Corrosion Society, Gdańsk (Poland), e-mail: sekretarz@psk.org.pl*
[**]*Road and Bridge Research Institute, Warsaw (Poland)*
[***]*The Łukasiewicz Research Network - Institute for Engineering of Polymer Materials and Dyes, Gliwice (Poland)*

Abstract:

Highly zinc-pigmented primers are one of the best primers dedicated to long-term corrosion protection of steel structures.

Taking into account the environmental aspects and sustainable development, it is important to develop new, more ecological products, such as waterbased zinc-pigmented primers, but with a reduced zinc content.

The paper presents the preliminary results of research on products based on specially selected, new and ecological formulations, carried out under the international project with the acronym EcoWaterZinc.

Keywords:

waterbased zinc-pigmented primers, ecological paint, primers with reduced zinc content

Implementation of additives from renewable sources in epoxy composites with lower negative environmental impact

Adrian Zajac[1], Dawid Zielinski[1,2], Andrea Szpecht[1], Marcin Smiglak[1]

[1]Poznan Science and Technology Park, Poznan, Poland
[2]Adam Mickiewicz University, Poznan, Poland
Email: adrian.zajac@ppnt.poznan.pl

In recent years, polymers from renewable sources are attracting a great deal of attention because of their possible sustainability, biodegradability and low toxicity. Moreover, it is time to think about limited petroleum resources and processes with controlled carbon dioxide emissions lead to global warming.[1]

Thus, there are numerous literature reports about the production of natural epoxy bio-resins. Some scientists are striving to produce epoxy resins using epichlorohydrin and bisphenol A, but largely derived from environmentally friendly processes. The development of these technologies is difficult, mainly for reasons related to the need to equalize production costs of conventional and ecological methods. On the other hand, many scientific groups are trying to obtain bio-resins from new raw or recycled materials: higher fatty acid esters, [2] cardanol [3] or vanillin. [4] However, the most interesting and the simplest way to reduce use of conventional raw materials is the implementation of natural additives. Those compounds can be added during epoxy resin production or directly with curing agent to make final resin-hardener system. Consequently, this operation allows to produce composites with up to 40% w/w content of bio-substances. In spite of many advantages it is possible to find difficulties, mainly with compatibility of ingredients.

However, in this work we present some natural bio-additives dedicated to epoxy resins, based on renewable resources. To increase compatibility with different resins we propose substances with epoxy-group and various structures. Proposed technology does not require any significant changes to the application process, but is based on molecules with epoxy groups and changes around it, finally replacing part of epoxy resin commercially used. In conclusion, we present a few implementations of our additives in various epoxy systems with amount from 10 up to 30% w/w.

The project is financed in frames of Intelligent Development Operational Program 2014-2020, Operation 4.4, "Innovation Incubator 4.0".

References:

[1] Thakur, V.K., Kessler, M.R. Biocomposites Composed Of Bio-Based Epoxy Resins, Bio-Based Polyphenols And Lignocellulosic Fibers. *Green Biorenewable Biocomposites: From Knowledge to Industrial Applications*. Apple Academic Press. 111 – 160, (2015)

[2] Baroncini, E.A., Yadav, S.K., Palmese, G.R., Stanzione, J.F. Recent advances in bio-based epoxy resins and bio-based epoxy curing agents. *Journal of Applied Polymer Science*. 133(45), 44103 – 44122, (2016).

[3] Atta, A.M., Al-Hodan, H.A., Hameed R.S.A., Ezzat A.O. Preparation of green cardanol-based epoxy and hardener as primer coatings for petroleum and gas steel in marine environment. *Progress in Organic Coatings*. 111, 283 – 293, (2017)

[4] Shibata, M., Ohkita, T. Fully biobased epoxy resin systems composed of a vanillin-derived epoxy resin and renewable phenolic hardeners. *European Polymer Journal*. 92, 165 – 173, (2017)

Synthesis of new oligoesterols based on waste poly(ethylene terephthalate) and dimerised fatty acids as a binder for polyurethane organic coatings

S. Waśkiewicz[a], P. Dziendzioł[a], E. Langer[b]

[a]Department of Physical Chemistry and Technology of Polymers, Faculty of Chemistry, Silesian University of Technology, 9 Strzody, 44-100 Gliwice, Poland, e-mail: sylwia.waskiewicz@polsl.pl

[b]The Łukasiewicz Research Network - Institute for Engineering of Polymer Materials and Dyes, ul. M. Skłodowskiej-Curie 55, Toruń 87-100, Poland

The goal of this scientific study is synthesis of new (oligo)estrols as a binder of polyurethane organic coatings. These oligoestrols are obtained by transesterification of waste poly(ethylene terephthalate) (PET) with derivatives of dimerised fatty acids (DFA) or fatty diols (FOH). The best properties as a binder have non-crystallizing compounds with appropriate viscosity. The viscosity of the binder is an important parameter because it determines how much volatile organic compounds content in varnish compositions can be reduced.

The proposed dimerized fatty acids (e.g. Pripols) have many properties that are desirable in protective coatings, namely: corrosion inhibition, elasticity, excellent moisture barrier, and thermal oxidation stability. In addition, the use of waste PET supports the packaging utilization program, which according to EU standards should be at the level of 65% by 2025 in member countries. Obtaining (oligo)estrols using chemical recycling from waste PET is one of the solutions for developing this material. According to literature data, including patents, the preparation of (oligo)estrols based on PET is mainly carried out by a glycolysis reaction. Glycol in the procedures described is often used in large excess in relation to the PET repeating unit, which leads to the first and foremost - terephthalic acid diester and a given glycol, including bis(hydroxyethylene terephthalate) as the repeating unit of the PET main chain, and unreacted glycol remains. These terephthalic acid diesters exhibit instability during storage at room temperature - the resulting glycolysate crystallizes over time, and thus its viscosity changes over time, and the reproductive characteristics of these oligoestrols in PU synthesis are unfavorable. In addition, the current free glycol in the mixture later contributes to the formation of irregular networks in polyurethane compositions. The use of DFA or FOH will ensure the flexibility of the oligoestrol chain, and their branched structure will limit the association of ethylene and terephthalate units, and thus the creation of crystal domains in the obtained oligoesters. In addition, scientific studies on polyurethane materials describe that polyurethanes from polyols which contain both a fatty dimer residue and an H-bonding group, can show self-healing properties.

References:

[1] Rajput, S.D., Hundiwale, D.G., Mahulikar, P.P., Gite, V.V. Fatty acids based transparent polyurethane films and coatings. *Progress in Organic Coatings. 77,* 1360–1368, (2014)

[2] Ahmad, M.U., Fatty acids: Chemistry, Synthesis, and Applications. *Elsevier.* (2017)

[3] Gogoi, G., Gogoi, S., Karak, N. Dimer acid based waterborne hyperbranched poly(ester amide) thermoset as a sustainable coating material. *Progress in Organic Coatings. 112*, 57–65, (2017).

Easy-to-apply chemical pre-treatment for long lasting bonding with typical adhesives on industrial polyolefins

G. Morgese[1], *R. Brönimann*[1], *J. Inauen*[1], *K. Siegmann*[1], *M. Winkler*[1]

[1]*Institute of Materials and Process Engineering (IMPE), School of Engineering (SoE), Zürich University of Applied Sciences (ZHAW), Technikumstrasse 9, CH-8401 Winterthur, Switzerland. giulia.morgese@zhaw.ch*

Gluing polyolefins (*e.g.* PE and PP) results in a very challenging task. The main reason for this evidence relies in their low surface energy, which reduces the affinity between the polyolefin surface and the chosen adhesive. To tackle this problem, the most commonly used solutions are physical surface treatments, such as plasma, corona and flame, which introduce hydrophilic moieties on the plastics surface, thus increasing their surface energy. These approaches require special setups, are unspecific and can induce material degradation [1]. Furthermore, they provide only a transient solution [2], making the storage of pre-treated substrates impossible. Due to all these drawbacks, there is the need for a robust alternative.

Here, we developed an easy-to-apply primer for long lasting bonding of adhesives on PE and PP. Our primer contains a surface-anchoring moiety and an adhesive-binding group to covalently react with the polyolefin substrate and with the glue. As surface-anchoring moiety we chose the perfluorophenyl azide (PFPA), which is known to undergo a C-H insertion reaction upon UV-activation [3], while as adhesive-binding groups we selected OH functions, which can covalently react with polyurethane or epoxy glues. We discovered that, when these two features (*i.e.* PFPA and OH) are combined in a single molecule, the reaction with the substrate does not occur and the molecule is only physisorbed, inducing no adhesion improvement. Chemisorption occurs only when the primer is a bicomponent formulation, comprising a hydrophobic tri-PFPA and a macromolecule bearing OH groups. The latter can be either commercially available, *i.e.* tannic acid, or synthetic, such as a polymer bearing simultaneously OH groups and PFPA.

In this work two different bicomponent formulations (tri-PFPA + tannic acid and tri-PFPA + Polymer-OH) were applied on PE and PP. After UV-activation, their adhesion performance with commercial (2k-epoxy) and industrial (PU-based) adhesives was tested and compared with untreated and plasma-activated references. All the formulations performed better than plasma on PE and PP with PU-based adhesives. With 2-k epoxy, a better performance than plasma was observed on PP, but not on PE. This latter evidence depends on the type of PE tested.

Since a good alternative to physical treatments should also be long lasting, UV-activated and plasma-treated substrates were stored at room temperature for up to 2 months before adhesion testing. Aged samples performed as good as freshly prepared samples with all the adhesives tested, while plasma partially lost its function overtime. Finally, the potential of the primers was tested on different adhesives, showing universality on PP.

[1] Shikova, T. G., Rybkin, V. V., Titov, V. A., Shutov, D. A., Kuval, E. V. *High Energy Chem.*, 39, 337 (2005)

[2] Guimond, S., Wertheimer, M. R. *J. Appl. Polym. Sci.*, 94, 1291 (2004)

[3] Siegmann, K., Inauen, J., Villamaina, D., Winkler, M. *Appl. Surf. Sci.*, 396, 672 (2017)

AUTHOR INDEX

Polish Association of Chemical Engineers
SITPChem Main Board
ul. Tadeusza Czackiego 3/5
00-043 Warsaw

CURRAN ASSOCIATES INC.
proceedings.com

ISBN 978-1-7138-7633-5